Ecotoxicology of Earthworms

General Editors
Dr P.W. Greig-Smith
Dr H. Becker
Dr P.J. Edwards
Dr F. Heimbach

Intercept

Published by Intercept Ltd, P.O. Box 716, Andover, Hants SP10 1YG, United Kingdom.

British Library Cataloguing in Publication Data
A CIP catalogue record for this book is available from the British Library

ISBN 0 946707 40 5

Typeset by J&L Composition Ltd, Filey, North Yorkshire

Printed by Athenaeum Press Ltd, Newcastle upon Tyne, UK

Contents

CHEMICAL BIOMONITORING WITH EARTHWORMS

POSTER PAPERS

RECOMMENDATIONS

Introduction

H. BECKER

Federal Biological Research Centre for Agriculture and Forestry, Königin-Luise-Strasse 19, D-1000 Berlin 33, Germany

There is a strong and growing interest in soil organisms among environmental scientists and regulators. This reflects concerns about soil fertility in agricultural land, risks of chemicals leaching into drinking water, contamination of soil, and its effects on plants and animals. Soil organisms can provide valuable indicators by which to study these issues. Recent developments in national and international legislation have sharpened the need for reliable, sensitive indicator organisms to use in research, monitoring, and regulatory testing.

Earthworms are good indicators of soil quality. They contribute to the complex processes of decomposition, to the degradation of plant debris and to the mineralization of organic materials. They mix the soil, improving aeration and water transport, and help to maintain soil structure.

Therefore, in the search for terrestrial organisms suitable to use in ecotoxicological testing of chemical substances in Europe, earthworms are an obvious choice. However, there are objections against the use of earthworms, in particular against Eisenia fetida as a standard test species because it is considered to be relatively insensitive to chemicals. This debate is important when earthworms are used widely as indicator organisms representing the whole of the soil environment.

Current regulations in several European countries demand data from tests on earthworms, carried out according to the OECD Guideline 207 and the corresponding EC test method. Further laboratory methods, for studying sublethal effects, and methods for standardized field tests, are being developed to extend the range of information available for the registration of chemicals, and to improve understanding of potential adverse effects.

Experience from tests performed with organisms exposed to chemicals via soil or artificial soil substrates provides evidence that it is the bioavailability of the substance which is the main factor determining the severity of effects on organisms. For a hazard assessment process evaluation of the exposure of soil-living organisms in the field is essential in addition to ecotoxicological tests performed in the laboratory. An internationally harmonized approach, incorporating ecological factors, the population dynamics of earthworm species, and interactions with other soil-living organisms is the best way to make this process successful. The goal should be to develop predictive models which estimate the probability that a chemical substance will affect soil organisms.

In order to exchange ten years experience in the field of earthworm ecotoxicology and to seek international agreement on these issues, an International Workshop on Earthworm Ecotoxicology was held from 10th to 12th April 1991, at the University of Sheffield, UK, in association with the Society of Environmental Toxicology and Chemistry's (SETAC) first European Conference.

At the invitation of the organising committee H. Becker (Federal Research Centre for Agriculture and Forestry (BBA), Germany) (Chairman), C. Callahan (Environmental Protection Agency, USA), C.A. Edwards (University of Ohio, USA), P.J. Edwards (ICI Agrochemicals, UK), P.W. Greig-Smith (MAFF, UK) and F. Heimbach (Bayer AG, Germany), 67 specialists contributed to an active programme of invited talks, posters and discussion sessions.

The workshop offered an opportunity to share experiences with the widely used OECD mortality test, to present results from field testing, to exchange ideas concerning new methods, sequential test programmes and risk assessment procedures, and to discuss ways of tackling the unsolved problems.

Some of the major issues that will benefit from the international cooperation fostered at this conference are:

- the selection of earthworm species as test organisms
- their significance as indicators of potential chemical effects on soil fauna
- the influence of sublethal effects on natural populations of earthworms and approaches to measuring such effects
- interpretation of the results of field tests in relation to natural population dynamics and other forms of human disturbance
- the triggers for further testing within a sequential assessment procedure, and the criteria to be used for classifying chemicals as dangerous for earthworms
- the improvement of knowledge of the ecology of earthworms in order to aid the interpretation of effects.

The presentations made by invited experts, and the posters offered by delegates are published in this volume. There is a strong need for internationally harmonized actions in order to avoid unnecessary testing and to promote the uniform acceptance of risk assessment decisions. Therefore a major aim was to formulate agreed recommendations based on the presentations and on the intensive, wide-ranging discussion sessions. The summary of these discussions and the recommendations prepared by Peter Greig-Smith are an important part of this volume.

I believe that these recommendations and the gaps in knowledge that have been identified will be a motivation for intensive work in the field of earthworm ecotoxicology and in the broader aspects of ecotoxicology of the soil fauna.

On behalf of the organising committee, I wish to thank the sponsors who generously provided the funds that allowed the Workshop to take place: BASF, Bayer, CIBA-Geigy, Duphar, Hoechst, ICI and Schering.

Prof. Peter Calow and Miss Julie Yeardley of Sheffield University gave us valuable organisational assistance. Many of the participants acted as referees for the papers in this volume, and I extend thanks to them all for their support.

Measuring the effects of chemicals on earthworms in the laboratory

A Review of Ecotoxicological Test Methods Using Earthworms

A.J. REINECKE

Department of Zoology, Stellenbosch University, South Africa

Abstract

A general review of laboratory tests for the assessment of the toxicity of chemicals to earthworms is presented. An evaluation of these methods based on the literature and experience is attempted stressing the advantages and disadvantages of each. Results of comparative tests are presented for the more general methods which are described fully and discussed. Toxicity testing methods intended for widespread use in environmental protection programs have to be sensitive and reproducible in predicting field hazards but should also be cheap, simple and fast to perform. Further research into test design and the use of models is required in order to obtain methods that are reliable and have good predictive value, especially in terms of the more subtle sublethal effects of environmental contaminants.

Introduction

Ever since the work of Darwin (1881) earthworms have been considered to be important in soil formation processes and in monitoring soil structure and fertility. They contribute by incorporating decaying organic matter into the soil, improving soil aeration, drainage and moisture holding capacity. It is therefore important to maintain a healthy earthworm population in soils and to avoid soils becoming contaminated.

Ecologically earthworms are near the bottom of the terrestrial trophic levels. Apart from being sensitive to some chemicals they tend to concentrate certain insecticides (Edwards & Thompson, 1973) and heavy metals (Carter *et al.* 1983; Ireland, 1983) in their body tissue. The concentrated chemicals may not harm the worm but will affect other organisms higher up in the food chain.

Since earthworms are essentially cosmopolitan and common in many soils they are vulnerable to impacts on soil. They are easily available and easy to handle and can be used as a bioindicator and test organism for chemicals in the soil environment.

Concern for the impact of toxic and hazardous substances in the soil environment has increased and efforts in developing methods for evaluation of these effects have correspondingly been stepped up.

The Commission of the European Communities concerning the registration of chemicals has made certain ecotoxicological tests mandatory since 1981. Ecological data on an algal test, a Daphnia test, a higher plant test, a fish test, a test for species accumulations, a biodegradation test and a test on an earthworm are now required before placing chemicals in certain quantities on the market.

The development of ecotoxicological evaluation procedures have come a long way but little progress has been made in designing tests that are simple, using readily available materials, are reproducible, cheap and are ecologically relevant to the soil situation. In practice acute toxicity tests are designed typically to determine the concentrations of a chemical which produces harmful responses on the test animal during continuous short term exposure. Often acute toxicity tests are equated with lethal tests aimed at determining LD_{50} or LC_{50}. But setting standards for environmental protection purposes is not that simple since delayed effects are not easily accounted for. The question of 'environmentally acceptable' concentrations of a chemical under conditions of chronic exposure cannot be answered from the results of acute lethal tests. The sublethal effects a chemical can have at low concentrations on behaviour and reproduction (Venter & Reinecke, 1988) are not revealed by LD_{50} and LC_{50} tests. Due to a lack of suitable sublethal tests standardization of lethal toxicity remains important in order to gain an understanding and compare the toxicity of various environmental contaminants.

This paper reviews the methods available, points out the questions still remaining to researchers and provides suggestions for future approaches.

Existing methods of testing toxicity

Methods have been reviewed by Satchell (1955), Davey (1963), Edwards & Thompson (1973), Edwards (1984), Heimbach (1988), and Goats & Edwards (1988) and are described in the OECD-guideline no. 207.

Some researchers have tested the effects of contaminants on earthworms kept in soils in containers in the laboratory (Edwards & Lofty, 1973; Hoy, 1955; Heungens, 1969; Caseley & Eno, 1966; Atlavinyte, 1975; Agarwal, 1978; Stringer & Wright, 1973; Kale & Krishnamoorthy, 1979; Lofs-Holmin, 1980 and Rhett *et al.* 1988). These tests tended to produce consistent and reproducible results because standard numbers of test earthworms of a single species are used and the worms are in intimate contact with the chemical. However, a wide range of soil types of greatly differing adsorptive properties have been used for these laboratory tests, and the LC_{50} for any particular chemical can differ greatly, depending on the test soil (Edwards, 1984).

Toxicity of chemicals to earthworms has been determined by immersing them in diluted solutions of chemicals for set periods of time and then transferring them to clean soil to assess mortality (Goffart, 1949; Martin & Wiggans, 1959; Edwards & Lofty, 1973; Stringer & Wright, 1973; Stenersen,

1979). Reproducible results can be obtained and an LC_{50} can be calculated for any species of earthworm. According to Lord *et al.* (1980) there often was a good correlation between the toxicity to earthworms of many chemicals in water and their toxicity in soil, but there were also many exceptions. This made it impossible to predict reliably whether a chemical which was very toxic to earthworms in an immersion test would be equally toxic or hazardous if it reached soil as a pollutant. Chemicals are adsorbed differentially onto the clay mineral and organic matter fractions in soils.

Assessment of toxicity of chemicals to earthworms by injecting them through the mouth (Stringer & Wright, 1976) or into the body cavity (Stenersen 1979) has also been attempted. According to Edwards (1984) and Goats & Edwards (1988) it is impossible to relate such toxicity data to field toxicity because this is a very artificial type of exposure to chemicals bearing little relationship to natural exposure in the field.

In the methods involving voluntary ingestion of contaminated food (Fayolle, 1979), food repellency affected mortality assessment. Stringer & Wright (1973) investigated the possibilities of feeding *Lumbricus terrestris* with poisoned leaf tissue. This method presented a difficulty in controlling the dose of chemical applied to the earthworms, since they consume different quantities of leaves.

Karnak & Hamelink (1982) used a carefully characterized and readily available soil type to run tests of chemicals against *Lumbricus terrestris*. A similar method of using a well-defined internationally common artificial soil (25% clay, 50% silt and 25% sand) has been adopted by the U.K. Pesticide Safety Precautions Scheme (P.S.P.S.) for pesticide registration purposes.

According to Edwards (1984) these methods suffered from various shortcomings and did not fulfil the requirements of a reliable ecotoxicity testing method with a high degree of reproducibility and relevance to the natural exposure of test organisms to a chemical in the field. This led to the development of various other tests, notably the simple filter paper contact test, the artifical soil test resembling loam soil and a test exposing earthworms to chemicals in a matrix of fine silica supported on glass balls (Ferreire *et al.*, 1981). The results of an inter-laboratory calibration (ring) test performed by 21 laboratories (Edwards, 1984) indicated that both the contact and artificial soil tests provided reproducible toxicity assessments for chemicals with an accuracy within an order of magnitude for most chemicals. The 'artisol' test seemed to give more variable results but was not conclusively evaluated.

A valuable comparison of the latter three methods proposed for EEC legislation is presented by Heimbach (1984, 1988). Since these three methods seem to be the most reliable at this stage a brief review of these methods is presented here with emphasis on the advantages and disadvantages in terms of reproducibility, environmental relevance and practicability.

A neurological assay assessing the conduction velocity of the giant axon of earthworms as related to concentration of a chemical was proposed by Drewes *et al.* (1983, 1988). This method has been proven reliable for determining sublethal neurophysiological effects and has great potential for

Table 1 A comparison of the LC_{50}-values obtained in the *Tubifex*-test after seven days and those obtained in the Artificial Soil Test (OECD) with *Eisenia fetida* after 28 days (Heimbach, pers. communication)

Test No	Substance Type	*Tubifex*-test mg/kg* (after Ammon, 1984)	Artificial Soil Test mg/kg* (OECD, 1983)
1	Insecticide	0.05 – 0.5	6 – 8
2	Fungicide	0.08 – 0.8	~ 496
3	Insecticide	0.09 – 0.9	1.0 – 1.8
4	Fungicide	0.50 – 1.5	17.4 – 26.7
5	Insecticide	0.2 – 2.0	~ 10.0
6	Fungicide	~ 2.5	~ 338
7	Fungicide	0.5 – 5.0	~ 626
8	Fungicide	0.7 – 7.0	~ 26.6
9	Insecticide	~ 4.2	~ 79.8
10	Fungicide	~ 4.5	~ 900
11	Ins., Herb.Fung.	1.0 – 10.0	~ 18 – 24
12	Fungicide	1.2 – 12.0	30
13	Insecticide	6.3 – 11.2	~ 6.7
14	Insecticide	2.0 – 20	20 – 200
15	Insecticide	20	20 – 112
16	Insecticide	4.3 – 4.3	~ 133
17	Insecticide	40	~ 3.6
18	Herbicide	10 – 100	> 1000
19	Herbicide	10 – 100	494 (424–575)
20	Herbicide	98 – 980	332 (302–315)
21	Herbicide	100 – 1000	917 (782–1162)
22	Herbicide	~ 850	> 300
23	Herbicide	1000 – 2000	> 1000

* a.i. = (active ingredient)

elucidating modes of toxic action. Its sophistication, however, limits its general use at this stage.

Several other methods have been proposed lately but have not been evaluated by other researchers. The Daniel-funnel test (Bieri *et al.*, 1989) showed promising results but needs wider evaluation. The Tubifex-test suggested by Bieri & Ammon (1990) has also been evaluated by Heimbach (personal communication) very recently. A comparison of this test with the artificial soil test (with *Eisenia fetida*) showed a poor correlation of 0·41 (see table 1 and figure 1). It is doubtful whether the Tubifex-test could be used successfully for general ecotoxicological evaluation of the soil environment.

Contact filter paper test (reference: OECD-guideline No. 207)

General: Single adult earthworms of the species *Eisenia fetida* are kept in glass vials lined with strips of standard filter paper treated with different concentrations of the test substance for 48 hours after which mortalities is determined.

Reference substance: Chloroacetamide

Test substrate: Standard quality filter paper (Whatman no 1, 80–85 gm^{-}22, medium grade containing the test substance and deionized water and (if necessary) organic solvents.

Control substrate: Standard filter paper and deionized water.

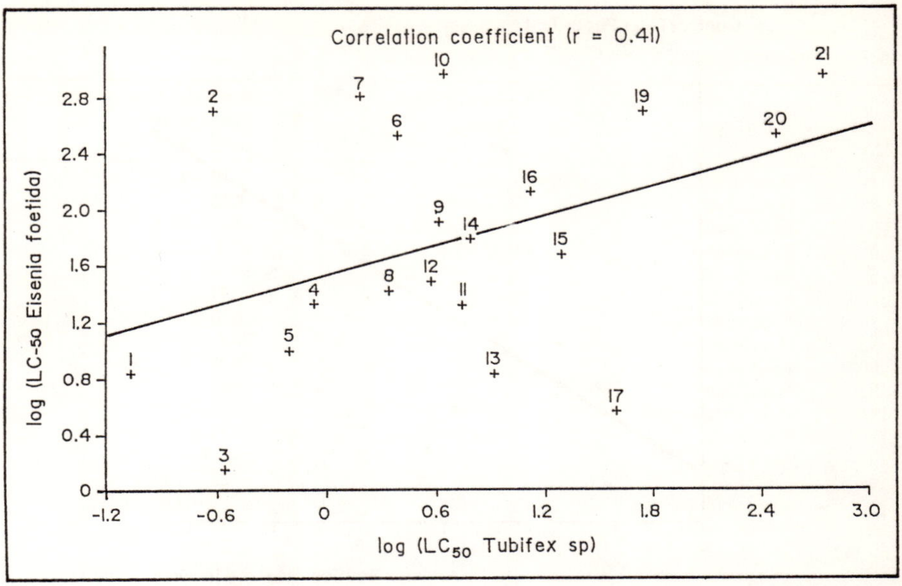

Figure 1. A comparison of the results obtained after testing 21 chemicals (table 1) with the Artificial Soil Test (with *Eisenia fetida*) and the Tubifex-test. A correlation coefficient of 0·41 was obtained. (Heimbach, pers. communication)

Containers: Glass vials 8 ± 1 cm long and 3 ± 0·5 cm diameter.

Test conditions: The vials are kept lying on their sides in a climatic chamber at 20 ± 2°C in continuous darkness

Test period: 48 hours

Test organisms: Adult *Eisenia fetida*, two months old clitellum with a biomass of 300–600 mg.

Test procedure: A range-finding test can be performed to provide information about the range of concentrations to be used in the definite test. Concentrations causing mortality of zero to 100% are determined. Concentrations are calculated in terms of test substance per cm^2 of filter paper starting from 1·0; 0·1; 0·01; 0·001; 0·0001 mg cm^{-2}.

Following the range finding test the definite test is performed by using at least five concentrations in a geometric series in order to find the LC_{50}. For each concentration, a control as well as ten replicates each with one worm per vial is used.

The test substance is applied (by pipette) in a 1 ml solution, emulsion or dispersion in deionized water or other solvent and evaporated to dryness. After drying 1 ml of deionized water is added to rewet the filter paper.

Earthworms are previously kept for a minimum of 3 hours on clean moist filter paper to evacuate the gut and then placed on the filter paper inside the vial. Vials are covered by plastic film on a perforated cap.

After 48 hours the numbers of dead worms are determined. Earthworms

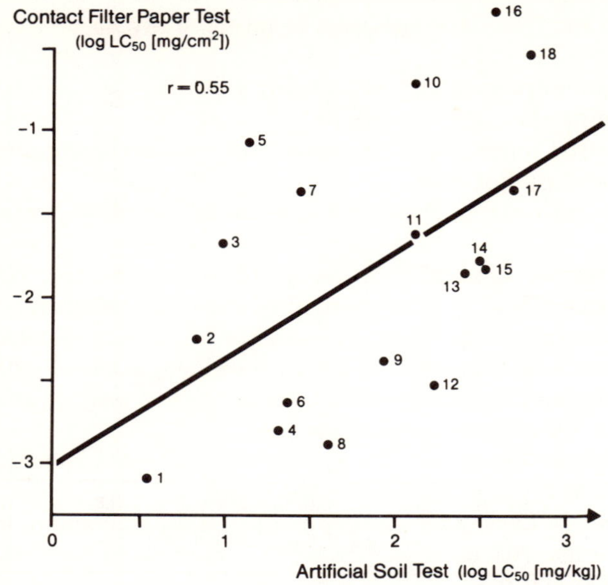

Figure 2. A comparison of the results obtained after testing 18 various chemicals with the Contact Filter Paper Test and the Artificial Soil Test with *Eisenia fetida*. A correlation coefficient of 0·55 was obtained (After Heimbach, 1988)

are considered dead if they do not respond to a gentle mechanical stimulus to the front end.

Presentation of results: Concentrations of the test substance with reference to the percentage mortality are reported. LC_{50} values (mg cm^{-2}) (confidence limits, $p = 0{\cdot}95$) are determined using standard methods (see Litchfield & Wilcoxon, 1949).

Evaluation: This test is easy to perform, but it gives only information about skin contact toxicity. Heimbach (1984, 1988) showed that there is little correlation between this method and the Artificial Soil Test (see figure 2). A correlation factor of 0·55 was obtained. Some chemicals which are highly toxic in the Contact Filter Paper Test are relatively non-toxic in the Artificial Soil Test and vice versa.

The test provides reproducible toxicity assessments for chemicals. It is an inexpensive screening procedure but is difficult to interpret in environmental terms. It can be used in conjunction with the Artificial Soil Test but probably has little ecotoxicological value apart from providing information on relative toxicity.

Artificial soil test (Reference: OECD-guideline No. 207)

General: Adult earthworms of the species *Eisenia fetida* are kept for 14 days in a defined artificial loam soil treated with different concentrations of the test substance and the mortality is determined.

Reference substance: Chloroacetamide.

Test substrate: The basic substrate in terms of dry weight) consists of:

- 10% sphagnum peat (about pH6) finely ground
- 20% kaolinite clay (›50% kaolinite)
- 69% industrial quartz sand (dominant fine sand with more than 50% of particle size 0·05–0·2 mm)
- 1% $CaCO_3$ pulverized, chemically pure to adjust pH if necessary.

The test substrate contains the basic substrate, the test substance and deionized water (25–42% of dry weight of the basic substrate).

Control substrate: It contains the basic substrate and deionized water.

Containers: Glass containers of about 1l capacity covered with plastic film or perforated plastic lids filled with an amount of wet test or control substrate equivalent to 500 g dry weight of substrate.

Test conditions: Containers are kept at 20 ± 2°C in a climatic chamber in continuous light (light intensity 400–800 lux).

Test period: 14 days (with an optional mortality assessment at day 7).

Test organisms: Adult *Eisenia fetida*, two months old with clitellum with a biomass of 300–600 mg.

Test procedure: A range-finding test can be performed to provide information about the range of concentrations to be used in the definite test. The substance should be tested at the following concentrations: 1000; 100; 10; 1; 0·1 mg per kg test substrate (dry weight).

One test batch per concentration and one for the untreated control, each with ten worms, are used in the range-finding test.

The definite test is performed using at least five concentrations in a geometric series to find the LC_{50}. At least four untreated controls and four test batches, each with ten worms per concentration are used. The results are presented as a mean standard deviation.

If insoluble in water, the test substance is dissolved in a small volume of suitable organic solvent (e.g. hexane, acetone or chloroform).

If the test substance is not soluble, dispersible or emulsifiable in organic solvents, 10 g of a mixture of fine ground quartz sand and a quantity of test substance necessary to treat 500 g dry weight of artificial soil are mixed with 490 g of dry weight of test substrate.

The test substrate must be ventilated before use. The amount of water evaporated must be replaced. The control should contain the same quantity of any additive agent.

For each test batch, an amount of wet test substrate equivalent to 500 g dry weight is placed into each glass container and 10 earthworms, which have been conditioned for 24 hours in a similar wet basic substrate, and then washed quickly and surplus water absorbed on filter paper before use, are placed on the surface of the test substrate.

Presentation of results: The concentrations of the substance tested are reported with reference to the corresponding percentages of dead earthworms. When the data are adequate, the LC_{50} (mg per kg dry weight) value and the confidence limits (p = 0·95) should be determined using standard methods (Litchfield & Wilcoxon, 1949).

Evaluation: The Artificial Soil Test was developed to overcome the variability between different soil types for a soil toxicity test. This test more accurately represents the natural environment of earthworms. Change in weight of worms can provide additional information about sublethal sensitivities of the worms. Results from this test correlate well with the 'Artisol' test (Heimbach, 1984, 1988). (See figure 3.) A correlation coefficient of 0·91 was obtained indicating a similar mode of uptake of the chemicals by the earthworms. A disadvantage of this test is that it is more labour intensive and time consuming and requires more laboratory space. It has the advantage that the exposure can be extended for longer periods of time provided that food is supplied. This allows for evaluation of sublethal effects after the standard test has been completed. Standardization is, however, still needed if long-term exposure is to be evaluated. The medium suggested by Bouwman & Reinecke (1990) could serve as a long term substrate.

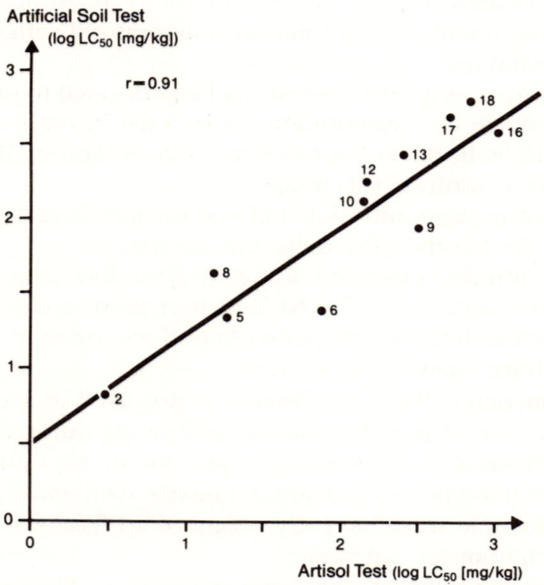

Figure 3. A comparison of the results obtained after testing 18 chemicals with the Contact Filter Paper Test and the Artisol Test with *Eisenia fetida*. A correlation coefficient of 0·48 was obtained. (After Heimbach, 1988)

Artisol Test (Edwards, 1984)

General: This test is quite similar to the Artificial Soil Test and requires adult earthworms of the species *Eisenia fetida* to be kept for 14 days on a basic artificial substrate called 'artisol' treated with different concentrations of the test substance. After 14 days of exposure mortality is determined.

Reference substance: Chloroacetamide

Test substrate: The earthworms are put into in a basic artificial substrate 'artisol'. The artisol is composed of:

- a skeleton of glass balls: diameter 1·5–2·0 cm: 1425 g of glass balls per container
- 90 g amorphous hydrated silica (trade mark 'Levilite') per container
- deionized water: 215 ml per test container

Glass containers about 1·5–2·0 l covered with perforated plastic film.

Test conditions: Containers are kept in a climatic chamber at a temperature of 20 ± 2°C in continuous darkness.

Test period: The test period is 14 days, at the end of which mortality is assessed.

Test organisms: Adult *Eisenia fetida*, two months old (with clitellums) and a biomass of 300–600 mg.

Test procedure: The concentration is expressed as the ratio between the test substance and dry weight silica. The support glass balls are not taken into account in the calculation.

A range-finding test can be performed to provide information about the range of concentrations to be used in the definitive test.

The substance should be tested at the following concentrations: 1000, 100, 10, 1; 0·1 mg test substance per kg dry weight silica.

One test batch with 10 earthworms per concentration and one for the untreated control is sufficient for the range-finding test.

At least five concentration steps are determined in a geometric series in the definite test so that the LC_{50} value may be found.

At least four untreated controls and four test batches, with 10 earthworms per concentration, are necessary. The results of these replicate batches are given as a mean with standard deviation.

The test substrate should be made up without any additional agents other than water. Immediately before starting the test, 215 ml of an emulsion or dispersion of the test substance in deionized water or other solvent is mixed with 90 g silica to obtain a homogeneous matrix.

If the test substance is not soluble, dispersible or emulsifiable, 10 g of a mixture of silica and a quantity of test substance necessary to treat 90 g dry weight of silica are mixed with 80 g silica and 215 ml deionized water.

If insoluble in water, the test substance can be dissolved in as small a volume as possible of a suitable solvent (e.g. hexane, acetone or chloroform).

The test substrate must be ventilated before use. The amount of water evaporated must be replaced. The control should contain the same quantity of any additive agent.

The matrix of silica, deionized water and test substance is mixed with 1425 g of glass balls and kneaded. This test substrate is placed in a test container and 10 earthworms (which have been washed and surplus water absorbed on filter paper) are placed onto the substrate.

Presentation of results: This is done similarly to that of the Artificial Soil Test mentioned previously.

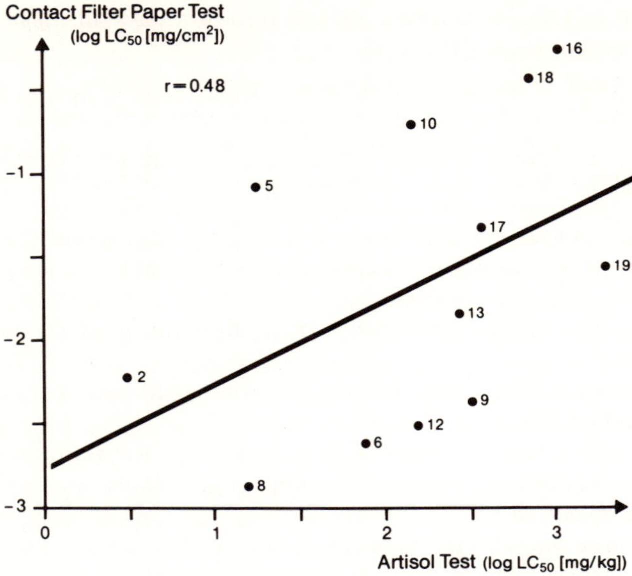

Figure 4. A comparison of the results obtained after testing 19 chemicals with the Artificial Soil Test and the Artisol Test with *Eisenia fetida*. A correlation coefficient of 0·91 was obtained. (After Heimbach, 1988)

Evaluation: The intercalibration exercise (Edwards, 1984) did not enable valid conclusions to be made since too few laboratories performed the test. The test is, however, very similar to the Artificial Soil Test since it more accurately represents the natural environment than the Contact Filter Paper Test. Results from this test correlate well with the Artificial Soil Test (Heimbach, 1984, 1988). (See also figure 3.) A correlation coefficient of 0·91 was obtained. The correlation with the Contact Filter Paper Test is very poor. (See figure 4.) A correlation coefficient of 0·48 was obtained. Experience showed that the 'Artisol' Test is easy to handle and to standardize and although the EEC has adopted the Artificial Soil Test as the method for investigating the toxicity of chemicals to earthworms (EEC, 1982, 1984) further evaluation and use of this method could be investigated.

Discussion

Earthworms can be used to evaluate the impact of contaminants by either measuring direct toxicity or by evaluating ecological impact. The methods described and discussed above are sufficient to determine relative toxicity of chemicals to earthworms but have limited ecological significance. As pointed out by Lofs-Holmin *et al.* (1988) laboratory tests can assess either lethal or sublethal effects. Sublethal tests are designed to detect subtle effects on the organisms such as disturbances in behaviour, retarded development, lowered fertility, teratogenic effects etc. which may cause population changes in the field although the animals do not show acute mortality. Data referring to the impairment of vital functions of organisms have a greater ecological relevance than those obtained from measurement of acute toxicity (Ma, 1984).

Ecotoxicological research is concerned specifically with the biological problems at the level of populations and communities. In this regard little information is available on the ways in which the influence of contaminants on predator–prey relationships, fecundity, changes in egg fertility, growth or behaviour could be utilized in ecotoxicological risk evaluations.

Furthermore, cheap and fast methods for measuring parameters are still lacking. Laboratory assays establishing at what level of stress the effects on earthworms may become demonstrable, are still not addressing the problem of assessing change in the ecosystem. The ecotoxicological approach is to characterize the ecosystem response to contamination and its subsequent recovery upon removal of the stress. In so far as the response of the individual earthworm could assist us in measuring or predicting the ecological change, laboratory assays will be useful – but only if a clear relationship can be demonstrated. We should, however, bear in mind that a suite of environmental indicators will be required with the earthworm representing only some facet of the ecosystem. The importance of the earthworm for the soil environment is stressed by the role it plays and its presence is a very important indication of the health of the system.

The science of ecotoxicology is largely based on extrapolation. Extrapolation has to be made from laboratory tests and microcosm studies to field situations across scales. Extrapolation requires valid models in order to make risk assessment meaningful. Without overemphasizing the role of models, their use must be recognized. In the designing of ecotoxicological tests we should therefore keep in mind that although the soil environment is highly complex, simple models could still be useful and we should develop simplified descriptions of the critical processes and components of the ecosystem. We have to know which species are crucial for maintaining system structure and function. We need to know the biology and basic responses of these species to be able to scale up from laboratory studies in order to predict ecotoxicological outcome in the field.

The next step in developing suitable ecotoxicological tests (and not mere toxicity tests) would be to show quantitative and valid relationships between responses obtained in laboratory tests and those observed in the environment.

The question remains: What should or could be measured to indicate the response of ecosystems to chemical stresses.

Computer assisted model evaluation of a variety of factors involved in the complex soil environment will probably provide the most reliable predictions.

References

AGARWAL, H.G.; YADAV, D.V.; PILLAI, M.K.K. (1978) Metabolism of the 14C-DDT in *Pheretima posthuma* and Effect of Pretreatment with DDT, Lindane and Dieldrin. Bulletin of Environmental Contamination and Toxicology 295–299.

AMMON, H.U. (1984) Worm toxicity tests using *Tubifex tubifex*. Compartement et effects secondaires des pesticides dans le sol. Versailles.

ATLAVINYTE, O. (1975) The effect of chemical substances on the activity of Lumbricidae in the process of straw disintegration. Proceedings of the 5th International Symposium of Soil Zoology. Ed. J. Vanek, Prague, Czechoslovakia. 515–519.

BOUWMAN, H.; REINECKE, A.J. (1990) A defined medium for the study of growth and reproduction of the earthworm *Eisenia fetida* (Oligochaeta). Biology and Fertility of Soils. Springer Verlag.

BIERI, M.; SCHWEIZER H.; CHRISTENSEN, K.; AND DANIEL, O. (1989) The effect of metaldehyde and methiocarb slug pellets on *Lumbricus terrestris* L. pp. 237–244. In: BCPC Monograph No. 41. Slug and Snails in World Agriculture.

CARTER, A.; KENNEY E.A.; GUTHRIE, T.F.; TIMMENGA, H. (1983) Heavy metals in earthworms in non-contaminated and contaminated agricultural soil from near Vancouver, Canada. In 'Earthworm ecology' J.E. Satchell (ed.) pp. 267–274.

CASELEY, J.C.; ENO, C.F. (1966) Survival and reproduction of two species of earthworms and a rotifer following herbicide treatments. *Proceedings of the Soil Science Society of America*, 30, 396–350.

DAVEY, S.P. (1963) Effects of chemicals on earthworms. A review of the Literature Special Scientific Report. *Wildlife No. 74, Washington*.

DARWIN, C. (1881) The formation of vegetable mould through the action of worms. Murray, London. 326 pp.

DREWES, C.D.; LANDA, K.B.; MCFAULL, J.L. (1983) Giant nerve fiber activity in intact, freely moving earthworms. *J. Exp. Biol.* 72, 217–227.

DREWES, C.D.; VINING, E.P.; CALLAHAN, C.A. (1988) Electrophysiological detection of sublethal neurotoxic effects in intact earthworms. In: Earthworms in waste and environmental management. C.A. Edwards and E.F. Neuhauser (eds.) 355–366.

EEC (1982) Directive 79/831, Annex V, Part C: Methods for the determination of ecotoxicity – level 1. Commission of European Communities, DG. X1/127–129/82.

EEC (1984) Directive 79/831 Annex V, Part C: Methods for the determination of ecotoxicity - level 1, Earthworms – artificial soil. Commission of the European Communities, DG, X1/128/82. Rev. 5.

EDWARDS, C.A. (1984) Report of the second stage in development of a standardized laboratory method for assessing the toxicity of chemical substances to earthworms. *Report of Commission of the European Communities. EUR 9360 EN,* 99.

EDWARDS, C.A.; LOFTY, J.R. (1972) Pesticides and earthworms. Report of Rothamsted Experimental Station for 1972. Part 1, 212.

EDWARDS, C.A.; THOMPSON, A.R. (1973) Pesticides and the soil fauna. Res. Rev. 45, 1–79.

FAYOLLE, L. (1979) Consequences of the impact of pollutants on earthworms III. Laboratory Tests. Documents Pedozoologiques. INRA Lab. Zooecol. Sol 1(1), 34–65.

FERREIRE, G.; FAYOLLE, L.; BOUCHE, M.B. (1981) Un Nouvel Outil, Essentiel pour l'ecophysiologie et l'ecotoxicologie: l'élevage les lombriciens en soil artificiel. *Pedobiologia* 22, 196–201.

GOATS, G.C.; EDWARDS, C.A. (1988) The prediction of field toxicity of chemicals to earthworms by laboratory methods. In: Earthworms in waste and environmental management. C.A. EDWARDS & E.F. NEUHAUSER (eds.) pp. 283–294 SPB Academic Publ. The Hague. Netherlands.

GOFFART, H. (1949) Die Wirkung neuer insektizider Mittel auf Regenwürmer. Anzeiger für Schädlingskunde 22, 72–74.

HEIMBACH, F. (1984) Correlation between three different test methods for determing the toxicity of chemicals to earthworms. Pestic. Sci. 15, 605–611.

HEIMBACH, F. (1988) A comparison of laboratory methods for toxicity testing with earthworms. In: Earthworms in waste and environmental management. C.A. EDWARDS & E.F. NEUHAUSER (eds.) pp. 329–335 SPB Academic Publ. The Hague. Netherlands.

HEUNGENS, A. (1969) L'Influence de la Fumure et des pesticides aldrine, carbaryl et DBCP sur la Faune du sol dans la culture des Azalees, *Revue Ecologie et Biologie du Sol.* 6(2), 131–145.

HOY, J.M. (1955) Toxicity of some hydrocarbon insecticides to earthworms. *New Zealand Journal of Science and Technology*. 367–370.

IRELAND, M.P. (1983) Heavy metal uptake and tissue distribution in earthworms. In: Earthworm Ecology, J.E. SATCHELL (ed.) pp. 247–265. Chapman & Hall, London.

KALE, R.D.; KRISHNAMOORTHY, R.V. (1979) Pesticidal effects of Sevin (Ci-naphthyl-n-methyl carbamate) on the survival and abundance of the earthworm Pentoscolex corethrurus. *Proceedings of the Indian Academy of Science.* 88B(1), 391–396.

KARNAK, R.E.; HAMELINK, J.L. (1982) Standardized method for determination of the acute toxicity of chemicals to earthworms. Ecotoxicol. Environ. Safety 6(2), 216–222.

LITCHFIELD, J.T.; WILCOXON, F. (1949) A simplified method of evaluating dose-effect experiments. *Journal of Pharmacology and Experimental Therapeutics,* 96, 99–113.

LORD, K.A.; BRIGGS, G.G.; NEALE, M.C.; MANLOVE, R. (1980) Uptake of pesticides from water and soil by earthworms. *Pesticide Science* 11, 401–408.

LOFS-HOLMIN, A. (1980) Measuring growth of earthworms as a method of testing sublethal toxicity of pesticides. *Swedish J. Agric. Res.* 10, 25–33.

LOFS-HOLMIN, A.; BOSTRÖM, U. (1988) The use of earthworms and other soil animals in pesticide testing. In: Earthworms in waste and environmental management. C.A. Edwards, & E.F. Neuhauser (eds.) pp. 303–313 SPB Academic Publ. The Hague, Netherlands.

MA, W. (1984) Sublethal Toxic Effects of Copper on Growth, Reproduction and Litter Breakdown Activity in the Earthworm *Lumbricus rubellus*, with Observations on the Influence of Temperature and Soil pH. Environmental Pollution (Series A) 33, 207–219.

MARTIN, L.W.; WIGGANS, S.C. (1959) The tolerance of earthworms to certain insecticides, herbicides and fertilizers. Oklahoma State University Experiment Station. *Processed Series*, 334.

RHETT, R.G.; SIMMERS, J.W.; LEE, C.R. (1988) *Eisenia fetida* used as a biomonitoring tool to predict the potential bioaccumulation of contaminants from contaminated dredged material. In: Earthworms in waste and environmental management. C.A. Edwards & E.F. Neuhauser (eds.), pp. 321–328 SPB Academic Publ. The Hague, Netherlands.

SATCHELL, J.E. (1955) The effects of BHC, DDT and Parathion on soil fauna. *Soils and Fertilizers* 18(4), 279–285.

STENERSEN, J. (1979) Action of pesticides on earthworms. Part I: The toxicity of cholinesterase-inhibiting insecticides to earthworms as evaluated by laboratory tests. *Pestic. Sci.* 10, 104–112.

STRINGER, A.; WRIGHT, M.A. (1973) The effect of Benomyl and some Related Compounds on *Lumbricus terrestris* and other Earthworms. *Pestic. Sci.* 4, 165–170.

STRINGER, A; WRIGHT, M.A. (1976) The toxicity of benomyl and some related 2–substituted benzimidazoles to the earthworm *Lumbricus terrestris*. *Pestic. Sci.* 7(5), 459–464.

VENTER, J.M.; REINECKE, A.J. (1988) Sublethal ecotoxicological effects of dieldrin on the earthworm *Eisenia foetida* (Oligochaeta). In: Earthworms in waste and environmental management. C.A. EDWARDS & E.F. NEUHAUSER (eds.) pp. 337–353 SPB Academic Publ. Hague, Netherlands.

Earthworm Species and Ecotoxicological Studies

M.B. BOUCHÉ

Laboratoire de zooécologie du sol (INRA), CEFE/CNRS, 1919 route de Mende, BP 5051, 34033 Montpellier Cedex, France

Abstract

Earthworms are an important food for vertebrates and have a major role in physical, chemical and biological processes in the soil. Laboratory studies on the toxicology of chemicals on earthworms are required to screen chemicals for any adverse affects of the chemicals on the earthworms and the ecological processes in which they are involved, before proceeding to more expensive field tests. The current morphological definitions of earthworm 'species' has resulted in the grouping of distinct non-interbreeding biological entities all together under the same name, e.g. Eisenia fetida complex and the Aporrectodea caliginosa complex. It is essential for laboratory tests that genetically homogeneous organisms are used such as biologically defined earthworm species, and/or that test animals are all derived from the same source. Although there is great taxonomic diversity among earthworms species, each can be characterised in terms of three ecological important categories, endoges, epigeics and aneciques.

The acute toxicity test must ensure intimate contact between the test chemical and earthworm by simulating natural soil but by using pure definable ingredients. Eisenia andrei (Bouché 1972) is recommended for this test.

The chronic toxicity tests should use a media that does not interfere with the interaction between the test chemical and the earthworm but allows cocoon production and growth to be assessed. A representative of each ecological category should be tested; endoge, Nicodrilus (= Aporrectodea) caliginosus caliginosus var paratypicus Bouché 1972; anecique, Nicodrilus (= Aporrectodea) caliginosus meridionalis Bouché 1972 ; epige, E. andrei. The last two species, which are more exposed to predators, are recommended for bioaccumulation studies.

To implement both the interpretation of the effect of chemical on earthworms and, through them, to agro-ecosystems and the acceptance of the use of chemicals, an initial analysis of the various steps to an integration of toxicological and ecological knowledges is given on the ground of the species field roles and the susceptibility to chemicals. The need, and the opportunity, today, to improve such an implementation is briefly discussed to demonstrate the absolute necessity to integrate researches to improve them and to serve practitioners.

Introduction

The increasing concern to provide a sound environmental assessment of chemicals requires that the interaction of chemicals and earthworms be taken into account. This was first realised after a general poisoning of the familiar american robin by way of their main food: earthworms.

While the study of earthworms was relatively underdeveloped, the ornithologists, among other environmentalists, focused attention to those tiny wriggly things as a key factor for wild life conservation. So, earthworms were one of the few organisms considered for ecotoxicological studies. Various acute toxicological tests were proposed and an easy-to-rear 'species': Eisenia fetida (Savigny) was chosen even though, this 'species' was known by some specialists as to be a species complex.

This first step – a pure artificial acute test – is a measure of the relative toxicity of chemicals. Obviously it does not provide sufficient information to predict the effect of a chemical applied in the field. So, an increasing effort has been made to improve field studies which are conducted on local earthworm communities of usually 3 to 10 species. The results from both laboratory and field studies can be grouped into three categories 1) acute toxicity tests (at first to establish an LC_{50} but providing other information at sublethal doses), 2) studies of chronic effects, 3) 'bioaccumulation estimations', i.e. chemical concentrations in earthworms, these earthworms being a vector in food chains (Figure 1).

All very well . . . but how to relate these results – gathered only for one (or a few) species in laboratories and for a few species in a few field conditions – to all earthworm communities, which hold thousands of species living in various soil, climate and management types? A method of interpretation and extrapolation of results is needed. Was it really necessary to spend so much effort to protect some birds?

In fact, birdwatchers were only looking at one role of earthworms, that of food for birds. Earthworms are a major component of the animal biomass of terrestrial ecosystems where they play several biological roles; as food for other organisms, interactions with plant roots and soil micro-organisms, chemical and physical functions which affect soil fertility and conservation. They are the main or occasional food of, 200 species of birds, mammals, amphibian and reptiles in France, plus many fish and invertebrates. They eat and crush an average of 300 Tonnes/ha of soil, mixing mineral layers and organic compounds to produce soil crumbs. The burrows they produce, provide channels through which soils respire. These channels also play a key role in infiltration of water allowing a mean through fall of 16 cm of water per hour (Al Addan *et al.*, 1991). Earthworms mediate many processes in the

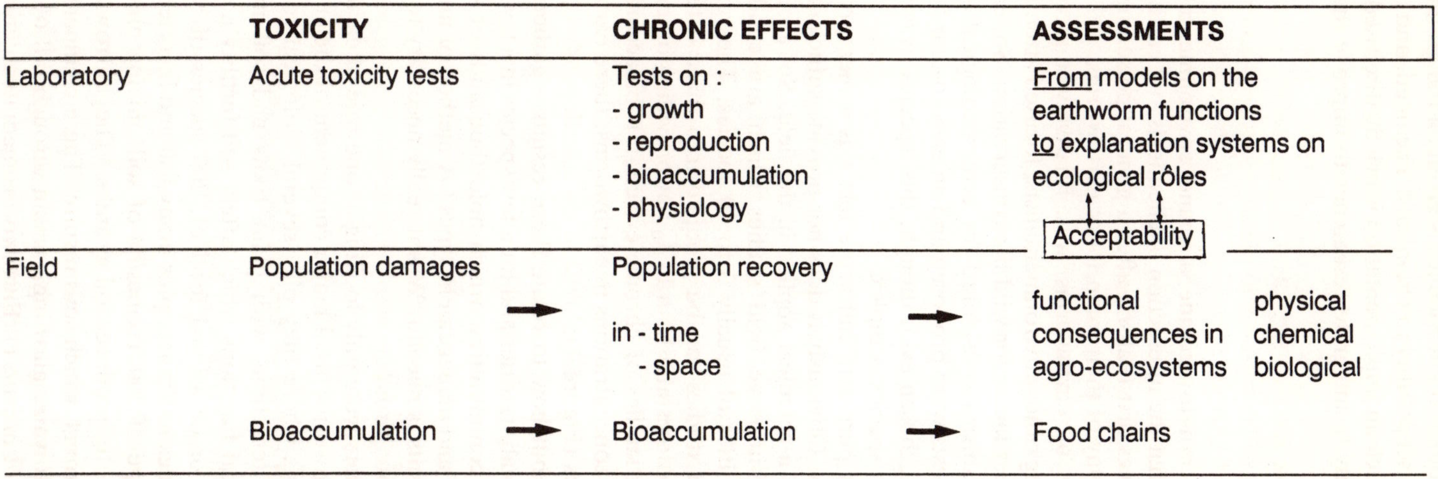

Figure 1. From the toxicity to the assessments. Contribution of both laboratory and field works to the appraisal of acceptable chemical effects.

nitrogen cycle. Direct effects include the huge gut transit (1200 kg N/ha/year) and excretion of small nitrogen molecules (ammonia, urea, microproteins: 400 kg N/ha/year) (Bouché *et al.*, in press), while their activities have indirect effects through changing soil microbial food conditions (Loquet *et al.*, 1977) and water logging.

So, we must thank birdwatchers who stimulated the protection of earthworms which in turn will contribute to better control of soil erosion, flooding, nitrate pollution and also an improvement of alternative methods such as minimum cultivation which need the earthworm task force as a substitute for most of the mechanical cultivation.

All earthworm species are not equally exposed to chemicals or are equally sensitive to them. However, in order to achieve a thorough environmental appraisal of chemicals in particular for the ecological processes involving earthworms, it is essential to realise that the effects of the chemicals on the environment is mediated through earthworm populations (Figure 2).

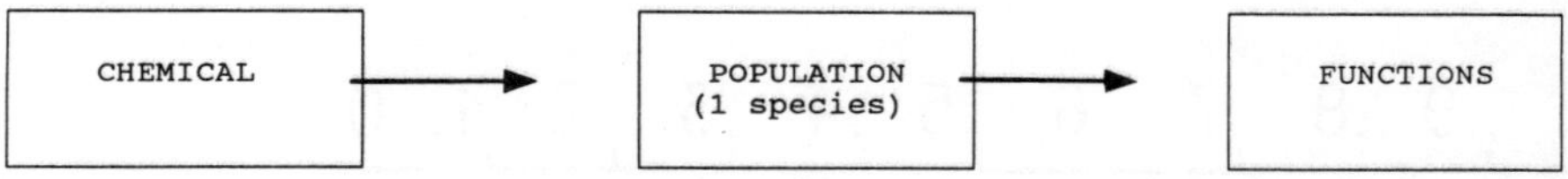

Figure 2. Relations between chemical input and ecosystemic function consequences.

The general validity of the model and its applicability depends on the functional position of the species of earthworms used in experiments, compared to the functional position of other earthworm species. This position depends, in turn, on the way of life (type of functioning) of the species, natural history and climate. There is an unavoidable need to extrapolate and integrate our knowledge from the few studies we have, both in toxicology and in functional ecology. Our ability to do this depends upon our skill to recognise species, and then to refer these entities to their roles. This in turn requires (i) our having a common concept of a species e.g. an entity composed of potentially interbreeding populations, (ii) our ability to distinguish between species and to recognise multispecies complexes.

Consequently, before we look which species to use in laboratory and field tests, a discussion of the functional position of a single species within the existing diversity is needed.

Functional position of species in their time-space diversity

PRESENT SPECIES AND TAXONOMY: THE STATE OF THE ART

The concept of a species as a group of potentially interbreeding populations can only be fully applied to a few groups of animals. For the rest, including earthworms, we have too few studies of the relevant properties.

There is a consensus by taxonomists on an earthworm classification based on those morphological characteristics which are preserved in museum specimens. However, earthworms bear relatively few morphological characters and it is therefore difficult to distinguish biologically defined species. On the other hand all stages of earthworms (gametes, eggs, juveniles and adults) have a very limited ability to move and therefore to mix their gene pools. Many other organisms e.g. ash trees are far more mobile due to their wind dispersed pollen and seeds. In other words we can suspect that taxonomists are being over-conservative in their concept of species and what is presently termed 'a species' based on morphology above could generally, if not always, be a mixture of true species.

For example, Eisenia fetida was recognized as heterogeneous since 1929 (Avel, 1929) and this fact, today well grounded, remains often ignored.

In this E. fetida complex, clearly Eisenia andrei is a good, relatively homogeneous, species while E. fetida is perhaps multispecific (Figure 3). E. andrei is a sibling species with at least one of the E. fetida 'species' (see Bouché *et al.*, 1988).

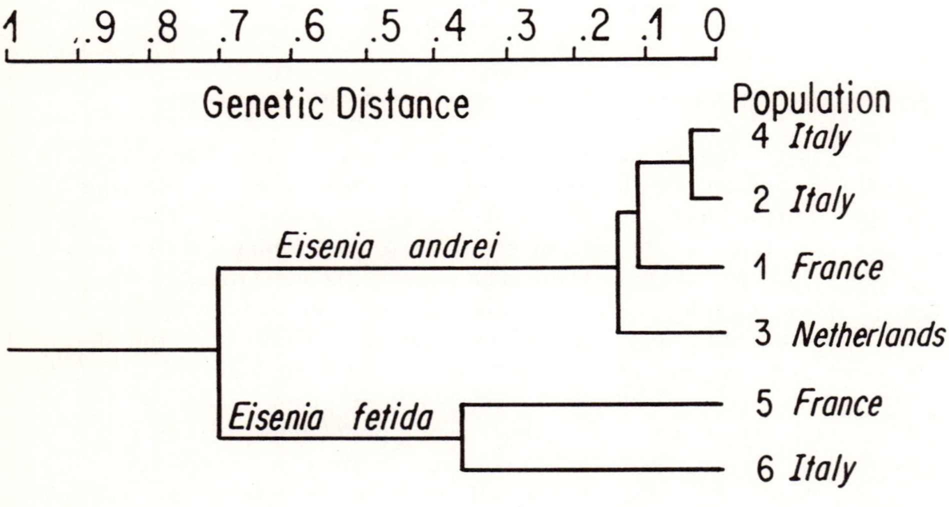

Figure 3. Genetic distances and diversity into the two sibling species of the Eisenia fetida complex (from Robotti, 1983).

An additional confusion here to non-taxonomists is which name to use for the genus. The question remains open as to whether A. caliginosa is a Nicodrilus (created by Bouché, 1972) or must be an Aporrectodea (having priority, created by Oerley in 1885) because we have no type and no original clear description of the species-type of the Aporrectodea genus. One 'species' (caliginosus) of this difficult-to-name genus is both very widespread around the world (Gates, 1972) and often plays a key ecological role.

About 50 species and subspecies of the genus have been described since 1826 when Savigny described Enterion caliginosa and Enterion terrestris. Taxonomists have recently been synonymising species based on preserved structures (see Easton, 1983) while an ecologist must recognise, or distinguish between entities with very different habits which live in the same field. For example Evans (1946) described A. nocturna, a big anécique species living in the same field in England as A. caliginosa, an endogé species but Easton (1983) synonymized these two totally different entities with no comment.

The A. caliginosa complex is multispecific, not totally described and the present 'specific' status of entities depends on taxonomists. However, this species complex is ecologically diverse and includes two definable entities, 'strains' that would be very suitable for laboratory tests on the chronic effects of chemicals (see below 'species and chronic tests') and should be relatively easy to rear. These two entities are A. caliginosa typica, an endogé and A. caliginosa meridionalis, a small anécique. In order to overcome problems with different interpretation of species definitions and genetic variation, well defined strains of each entity should be used.

THE HIGH SPATIAL AND TAXONOMIC DIVERSITY AND THE PRESENT VERSUS THE PAST

The colonization of terrestrial lands could be described since the primary era to the present (Bouché, 1983; Bouché, in press). The long lasting adaptation of earthworms to terrestrial conditions has led to an animal group with two main properties – 1) a very high biomass in many terrestrial soils, often earthworms are the major animal biomass in a soil, 2) a relatively poor ability to migrate with the consequence that selection of characters to suit the local environment is not diluted by the flow of genes between populations. With the result that some 'species' are mosaics of isolated populations.

The result is a key ecosystemic role and great taxonomic diversity. The great taxonomic diversity makes it very difficult to apply the results of laboratory or field experiments to all situations and we must even be cautious when applying them to the relatively few species that colonised areas of middle and northern Europe that were depopulated by the quaternary glaciations. As an example the 'well known' Lumbricus terrestris L. can be split in two non-interbreeding populations (two local 'species') in the same field (Figure 4).

Even though the taxonomic background is both complex and poorly known and field populations consist of mostly local subsets of 'species', we have fortunately a tool which gives us the chance to extrapolate the information on the faunal changes resulting from chemical uses between species, field and ecological consequences.

SPECIES PROPERTY CONVERGENCES AND ECOLOGICAL TYPES, OR HOW TO EXTRAPOLATE RESULTS BETWEEN COMMUNITIES VERSUS TIME-SPACE

Terrestrial ecosystems have several general functional features. Plants absorb most of their nutrient and water from the soil, and produce most of their

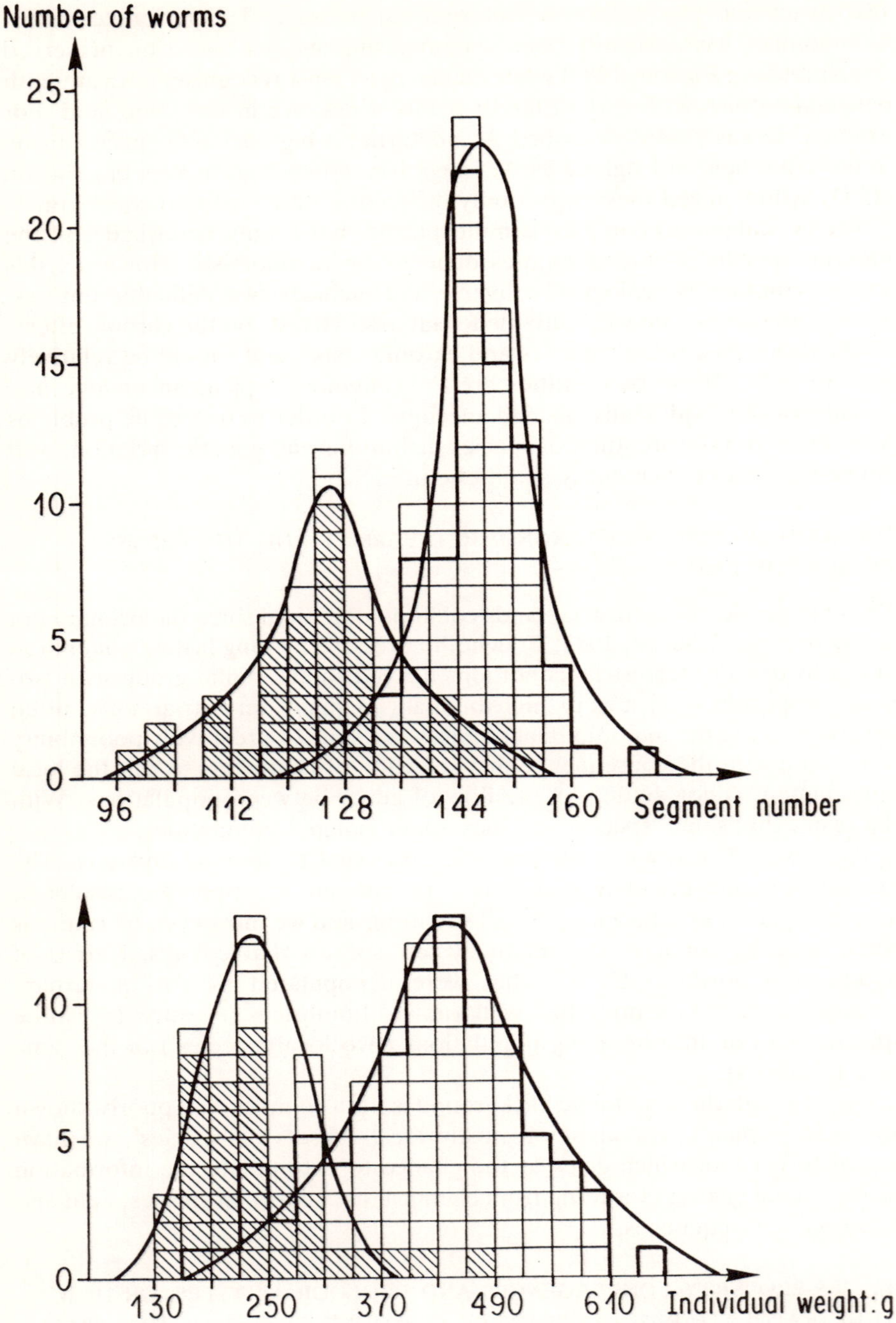

Figure 4. Observation of two distinct, most probably not interbreeding, amphimictic populations of the so-called Lumbricus terrestris L. (from Bouché, 1972). Is L. terrestris (at least) two species ?

(dead) tissues above ground. To resist herbivores, parasites and plants diseases, these tissues often include repellent factors such as alkaloids, resins, tannins, hard and poorly digestible tissues, spines.

The dead organic compounds from plants, directly or indirectly (faeces of decomposers or dead decomposers respectively), accumulate on the soil as litter. The decomposition of this litter depends on both the climate and the 'repellent quality' of its components. So that often, in bad seasons (dry or cold) or poor soils (where plants are repellent because more 'resistant') the litter is only slowly recycled and accumulates as an organic layer on the soil mineral horizon. This general pattern occurs more or less everywhere so that the various earthworm populations have evolved adaptations to enable them to recycle litter. This evolution has led to similar morpho-physiological characters in species of very different origins (family, genus) but converging on the same ecological roles and adaptations.

Three main ecological categories based on earthworm features (and not their milieu) were described (Bouché 1972) and could be recognized on the following features.

a) The endogés – or endogeics – are earthworms without skin pigmentation (i.e. flesh-tint more or less bloody or brown-earth by transparency) having an ecophysiological regulation of activity closely linked to soil conditions i.e. a resting stage or period of quiescence (hibernation or aestivation directly mediated by temperature and moisture). Great diversity of size, peripharyngial musculature, nephridial and circulatory adaptations occur in this group, depending of their sub-specialisation to feed on relatively organic rich food (polyhumics, carnivores) or poor food (oligohumics eating a lot of mineral soil), or living in well aerated or poorly aerated soils . . .

b) Epigeics – or epigés – are homochromically skin pigmented i.e. having a colour similar to their 'normal' milieu such as red-brown pigments in litter or sometimes green as in grasses because they are exposed to the selective presence of seeing predators. Often in the topsoil, they are destroyed by predation and physical factors as drought, frost or fire. They compensate these losses by high reproductive rate (r-selected: i.e. small size and a high number of hatchlings) associated with their rich organic food and by their 'resting' form as cocoons.

c) Anéciques – or anecics – are earthworms adapted to two milieus: they feed nightly at the soil surface on relatively fresh litter (where they later deposit their faeces) and they digest and regulate their physiology by precisely positioning themselves in soil gradients (1 to 6 m deep). They are subjected to night predators when they feed at the soil surface, their skin is darkly pigmented (brown-black or red-brown) especially the anterior and dorsal parts. Dragging their food into their burrows after an exploration of the available diet around these burrows (the tail remains in the burrow mouth), digging deep burrows to get of a greater choice of moisture and temperature in soil gradients, these earthworms open vertical channels in which they live most of the time (Bouché, 1987). To do that they are relatively large (7 cm to more than one meter in length), more strong musculature and a big gizzard able to mix the organic remains from litter with the mineral soil.

Under middle European conditions, the most stressful period for them is June when few plant residues remain as food, the short nights allow less time to eat, often there is drought and a high metabolic need because temperatures are high. These animals are typically adapted to these conditions by a specific resting stage: diapause. The diapause is mediated by photoperiod (Bouché, 1984; Saussey and Debout 1984) so that in June and mid-July they are resting irrespective of soil conditions. In the wetter and cooler parts of Europe an intermediate subcategory of earthworm – epianecics – occurs (i.e. Lumbricus terrestris, L. friendi, L. festivus, L. badensis, L. polyphemus . . .). This group differs from anecics by their lack of diapause. They always live in soils which remain moist.

The distribution of each of these ecological categories is determined by present soil and climate conditions and the recolonisation of areas following paleogeographic events (Bouché, 1983). Such historic, pedologic and climatic conditions must be carefully interpreted before results from ecotoxicological field tests can be generalised and applied in an European, or even worldwide context.

In other words laboratory tests are not directly relevant to field nor natural populations and field tests are not relevant to any time-space generalisation without a mechanism for extrapolation. The 'ecological categories' provide the mechanism which allows extrapolation both between species and between field populations. Let me demonstrate how to handle such concepts in practice.

Species and laboratory tests

SPECIES AND ACUTE TOXICITY SCREENING TESTS: ACCURACY AND REPRODUCIBILITY

Acute toxicity laboratory tests are an attempt to categorise chemicals into 3 groups; 'very toxic', 'non toxic' or 'intermediate toxicity'. The very toxic compounds must be discarded and the non toxic compounds could be accepted. While the chemicals of intermediate toxicity must be assessed by studies (chronic laboratory tests and field tests). Laboratory tests are just models, i.e. in any way they are substitute of true natural ecosystems where the chemical will be spread. To conclude to 'non toxic' or an absolute toxicity we accept that the relation earthworm/chemical is so obvious that the model could not be wrong versus real systems even if it is a harsh representation of these systems. We have absolutely no means to compare such models to the infinite number of ecosystem properties and all arguments to relate them to real soils are just arbitrary.

Nevertheless, the model must be reproducible to be accepted both as a useful tool and internationally. Three aspects of reproducibility are discussed.

1. An acute lab-test must ensure a close relation between the tested chemical and the earthworms, i.e. must eliminate interferences between this relation and the 'soil' as a substrate. A 'neutral' soil substrate is

required so that the 'soil' does not interfere with the interaction between the chemical and earthworm.

2. An acute lab-test must be reproducible, i.e. based on 'pure' (similar) components. All components ('soil' and ingredients, earthworm, and chemical) must be clearly defined. The more closely this is achieved, the greater the reproducibility. Even if only one component is not precise (as, for example, peat which varies with origin and with level in the same peat-bog) the reproducibility of a test may depend on it.
3. In order to satisfy points 1 and 2 above it is not possible to reproduce at all ecosystems, where soils are interacting with chemicals and where components are not pure and vary always in time-space. The so-called similarities of microcosms with filed conditions are just arbitrary opinions.

Further discussion about how to achieve neutrality of soil and reproducibility of the milieu and chemical in tests must be made elsewhere. A suitable species for an acute toxicity screening test must be easy to rear and be genetically homogenous. A species that fits these criteria is Eisenia andrei. It is genetically more homogenous than the E. fetida complex (see above 'Present species and taxonomy'). A further improvement would be to select a genetically controlled single strain of E. andrei which could be distributed to all laboratories.

SPECIES AND CHRONIC TOXICITY TESTS

Chronic tests are made to detect physiological disorders at sublethal doses which could occur in field. In addition to the criteria for lab tests described above, the milieu (environment) must include an earthworm food which will not interfere with the chemical tested. At present such a synthetic food is not available. The feeding behaviour of each of the three main ecological categories of worms can be affected by the type of food and the type of chemical.

Ideally a chronic test should include representatives of each ecological type of earthworm. E. andrei is a suitable epige. Endogeics and anecics are less easy to rear. However, suitable and definable, entities exist within the Nicodrilus caliginosus complex. I propose

1. to establish laboratory cultures of one strain of an endogé such as Nicodrilus caliginosus caliginosus var. paratypicus and one strain of a small typical anecic such as Nicodrilus caliginosus meridionalis,
2. to restrict toxicological laboratory tests to these strains of the selected entities, which include E. andrei, so that genetic consistency and homogeneity is maintained. These strains would be distributed to laboratories rearing live animals for tests,
3. to improve the definitions of the milieu for such tests: a similar milieu for epigé, endogé and anéciques; similar reproducible food; special care for photoperiodic reactions of the anecic strain ; easy to handle soil to follow chronic effects such as cocoon production or growth changes . . .

SPECIES AND BIOACCUMULATION TESTS

The conditions for studying the accumulation of chemicals in earthworms are very close of those described above. Nevertheless the need to kill the worms at intervals in order to follow the bioaccumulation pattern means that many more earthworms are required and therefore increased mass cultures. The greatest concern is to predict risks for birds and most mammals (exception moles, which eat endogeics) so these laboratory tests could be made with an epigé, (E. andrei) and anecic N. meridionalis. Again special care must be taken with food-chemical interactions which could affect both food selection in the milieu and changes in palatability.

Interpretation and acceptability

Having obtained results from laboratory and field tests both about acute and chronic demographic effects on populations and about chemical body burdens, and bearing in mind that earthworms are an important food source, what level of population disturbance and chemical burden can we accept?

Obviously, we cannot give an answer without taking in account:

1. The fact that chemical use is linked with other human activities (agriculture, industry, . . .). The protection of earthworms must be justified in comparison with other socio-economic factors.
2. The fact that environmentalists and more generally society, could not accept a decision without the demonstration of a good assessment, which uses state-of-the-art technology.
3. To make such an assessment we need to shift our primary interest from the relationship of the chemical and earthworm, to the integrated ecosystemic and economic role of earthworms. In other words toxicological and ecotoxicological studies are useless without a socio-economic and ecological evaluation.

This shift is possible only if we solve some problems caused by the weaknesses in our agro-ecological knowledge and of the absence of a really efficient integration between toxicology, earthworm ecology, agronomy, economics. Before discussing further the new possibility of forming our decisions using such an integration of information, I must point out one weakness in practice and the relations between species and acceptable perturbances.

SPECIES AND RESEARCHERS

While Europe has the highest concentration of earthworm specialists they are only the equivalent of one to three full time taxonomists on the whole continent. The various researchers with some skill in this matter are also involved in other activities and are not in a position to make a very classical work based on both comparative morphology and the use of modern techniques which are essential if we want to know on what species we work.

Not only does the general ecologist or toxicologist have no modern taxonomists available to help them but often there are no good text books on their local fauna. In consequence, researchers studying earthworms often use very doubtful or imprecise names for the 'species' on which they worked. The unreliable names used in reports and papers prevent a sound integration of knowledge to assist in the interpretation of the effect of chemicals on earthworms. We are now paying for a lack of earthworm researchers and the lack of general studies about the main animal component of our terrestrial ecosystems. We must fill this gap in our knowledge by assumptions!

SPECIES AND ACCEPTABLE PERTURBATIONS

Is a reduction of earthworm populations by 50% and the subsequent reduced population during the period of recovery, acceptable?

This simple and practical question includes three questions

1. What is the length of time required for the population of each species to recover to its original level?
2. Is the time for full recovery less than the time before other human action again reduces the population?
3. During the period(s) of reduced population is the decreased activities of the various earthworms in their physical, chemical and biological agro-ecosystemic roles acceptable?

Clearly to give a sound answer to the first question we need a demographic, an environmental and a functional appraisal.

DEMOGRAPHY AND SPECIES RECOVERY

There are cases where earthworm populations have substantially declined, in mining spoils or in moderately polluted soils for example. Heavy metals eliminate the susceptible genus Scherotheca in South France (Abdul Rida, com. pers.). Such situations are clearly unacceptable.

The acceptability of a partial and temporary poisoning depends on the time needed by a population to recover. We have at present no field studies to predict such a recovery but we can nevertheless build a predictive model based on the sparse knowledge we have from laboratory cultures and on the 'activity submodel' which relate the intensity of metabolism and the activity to three main driving factors: soil temperature and moisture, and photoperiod.

It is possible to make some generalisation on the rate of recovery of a species based on its size and the ecological functional type. The smaller the worm and more like an epige, the more likely a species is r-selected and has a high rate of reproduction and a short time from egg to maturity (2–5 months – depending on environmental factors). On the other hand the larger the worm and more it is K-selected, the slower the rate of population recover will be. Even though there are a large number of European earthworm species we can fortunately place those species in their ecological categories and apply our rare knowledge to both internal properties (genetical aptitude in the gradient of prolificacy: K to r) and on external regulations (physical environmental factors and food availability) based on their relative position in relation to the ecological categories. We need to build such an interspecific demographic model, to validate its predictability in the field and to use it.

SPECIES RECOVERY AND SUCCESSION OF POPULATION REDUCTIONS

To determine the acceptability of one population reduction resulting from the use of one pesticide we need to predict the time of recovery of the populations of earthworm communities and to place it in the context of other population reducing actions such as ploughing and other chemical uses during the year.

The frequency of such population reducing actions on earthworm is variable: very scarce in forests and permanent grasslands, these actions increase with the intensification in cropping (in orchards, in cereal crops, . . .). Acceptability of a chemical application depends on integrated assessment of the successive population reductions versus the ability of populations to recover.

SPECIES ROLES AND ACCEPTABILITY

When earthworm populations are reduced one or more times their ecological actions are reduced in proportion to their lower biomass.

The acceptability of reduced populations should not be fixed at an arbitrary level of decrease (such as 50%) but must result from an assessment of the various consequences of such a decrease for the ecosystem concerned. Examples of possible consequences to be assessed are:

(i) an appraisal of the loss of soil physical properties must be made in terms of a decrease in soil formation and stabilisation of granules, of water infiltration through- burrows, of increased risk of erosion,
(ii) an appraisal of the change in fates and movement of plant nutrients must be made as the balance of chemical elements in biomasses, necromasses, soil catchment and leaching, and seasonal synchronisation between compartments (as soil supply and plant needs).
(iii) biological consequences such as root distribution or microbial functions must be tackled. Especially, the consequence on earthworm predators (as fox, badger, woodcock, american robin, . . .) must be estimated both because the decrease of an important source of winter protein and because of the potential contamination of the remaining earthworms.

A new and integrated assessment is a prerequisite to general acceptance

INTEGRATED KNOWLEDGE AND ACCEPTABILITY

The use of chemicals on agro-ecosystems have been both an economical need and a blind action because the consequences of these uses are not fully predictable. Only some empirical properties have been observed while most, including all indirect effects were ignored.

Today this situation is changing dramatically both because we have a greater concern with a true agro-ecological assessment and mostly because we have recently gained the ability to do such a true assessment (Bouché, 1990). This change paves the way for linking preliminary laboratory tests, which discard or select useful chemicals, field tests, which measure the true chemical effects on populations and integrated assessments, which give a sound appraisal of the acceptability of a given population reduction versus the economic value of a given chemical use.

This integrated assessment needs the coordination of European research to aggregate all the available knowledge. This is possible thanks to two main tools, plus a conceptual transdisciplinary improvement.

1. Relational Data Bases, which are now reaching a new step (Distributed Relational Data Base), are computerised systems which manage all available data and automatically give access to the required data without the need of programming. The question only has to be asked. Thanks to these tools, facts (without pre-interpretations) will be available on the European scale.
2. Artificial Intelligence (A.I.) systems such as hypertexts, made possible for each specialist to produce the 'best' interpretation procedures; for example, to describe detailed models, to give their limits, their level of probability, their power to predict. Such A.I. systems could be organised in a matrix structure freely available to people and able to give a general interpretation and, if needed, to dig deeper to more detailed levels. All models on the effect of the chemical distribution technique, on earthworm exposure to the chemical in the field, on earthworm susceptibility versus chemical or solvent properties, on demographic population recovery, on effects of seasonal variation of temperature, moisture and light, on physical soil consequences, on transfer through food webs, . . . could be described by available specialists. These A.I. systems will be up-to-date and give state-of-the-art information. Improvement could be made every day on such systems. There is fossilization in papers.
3. These two tools need a very restricted number of transdisciplinary concepts which make it possible to handle the so-called complexity. These few concepts link data handling, interpretations, validations to field, . . . and could be easily translated between the various European languages. For example the 'use of pesticides' needs to be understood early by researchers developing new chemicals, by agronomists testing them and by regulatory advisers giving authorisations. In addition this knowledge must be available to practitioners or farmer advisers, to choose the 'best' available chemical based on the full facts. In such 'explained' situations there is greater acceptability because of the improved transparency from very specialised works to general uses. Because each specialist 'explains' their knowledge and makes it available, the use of these A.I. tools improves 1) the clarification of ideas, 2) the

availability of such ideas, 3) the transdisciplinary improvement of interpretation i.e. the synergy (for more details, see Bouché, 1990). The clarification eases translations between European languages and improves cooperation and availability to practitioners.

Last, but not least because of the vast number of European earthworm species, of the great variety of local conditions and of the increasing number of available chemicals it is essential to integrate our knowledge. The case is even more necessary when we consider the needs of the rest of the world. Only this integration paves the way to use both the various earthworm species for tests and the numerous chemicals in a sound, practical basis.

References

AL ADDAN, F.; ALIAGA, R.; BOUCHÉ, M.B. (1991) Relations entre peuplements lombriciens et propriétés des sols méditerranéens. In: Advances in management and conservation of soil fauna. VEERESH, G.K.; RAJAGOPAL, D.; VIRAKTAMATH, C.A. (eds). Vedams Books International, New Delhi (Inde).

AVEL, M. (1929) Recherches expérimentales sur les caractères sexuels somatiques des Lombriciens. *Bull. biol. Fr. Belg.,* 63, 149–318.

BOUCHÉ, M.B. (1972) Lombriciens de France. Ecologie et Systématique. Ed. *INRA, Ann. zool.-écol. anim., n. sp.,* 72–2, 1–671.

BOUCHÉ, M.B. (1983) The establishment of earthworm communities. In: J.E. Satchell (ed.) Earthworm ecology from Darwin to vermiculture. Chapman and hall London, pp. 431–448.

BOUCHÉ, M.B. (1984) Les vers de terre. La Recherche, 156, juin 1984, 796–804.

BOUCHÉ, M.B. (1987) The subterranean behaviour of the earthworms. In: A.M. Bonvicini Pagliari et P. Omodeo (eds) On earthworms, Selected Symposia and Monographs, UZI, 2, Mucchi Modena, pp. 159–169.

BOUCHÉ, M.B. (1990) Ecologie opérationnelle assistée par ordinateur. Masson éd. Paris, 572 pp.

BOUCHÉ, M.B. (in press) Les annélides lombriciens: une ré-évaluation. In: Decu (réd) Traité de zoologie du sol. Bucarest.

BOUCHÉ, M.B.; AL ADDAN, F.; CORTEZ, J.; HEIDET, J.CH.; FERRIÈRE, G.; MAZAUD, D. (in press) The earthworm role in the actual nitrogen cycle.

BOUCHÉ, M.B.; HAIMI, J.; HUHTA, V. (1988) Two earthworm taxa (Oligochaeta, Lumbricidae) new to Finland. Memoranda Soc. *Fauna flora Fennica,* 64, 65–67.

EASTON, E.G. (1983) A guide to valid names of Lumbricidae (Oligochaeta). In J.E. Satchell (ed.) Earthworm ecology from Darwin to vermiculture. Chapman and Hall, London, U.K., pp. 475–485.

EVANS, A.C. (1946) A new species of Earthworm of the genus Allolobophora. *Ann. mag. hist. ser. II,* 13, 98–101.

GATES, G.E. (1972) Contributions to North American earthworms (Annelida: Oligochaeta). III. Towards a revision of the earthworm family Lumbriciae. IV. The trapezoides species group. *Bulletin of Tall Timbers Research Station,* 12, 1–146.

LOQUET, M.; BOUCHÉ, M.B.; BHATNAGAR, T.; ROUELLE, J. (1977) Essai d'estimation de l'influence écologique des lombriciens sur les micro-organismes. *Pedobiologia,* 17, 6, 400–417.

OERLEY, L. (1885) A palaearktikus övben élö Terricolaknak revizioja és elterjedése. *Ertek. Term. tud. Kör.,* 15, 1–34.

ROBOTTI, C.A. (1983) Genetic distances among European populations of Eisenia fetida andrei and Eisenia fetida fetida. *Att. assoc. genet. Ital.,* 29, 207–208.

SAUSSEY, M.; DEBOUT, G. (1984) Nouvelles données sur le déterminisme de la diapause de Nicodrilus giardi (Ribancourt) (Oligochète, Lombricien). *C.R. acad. sci. , Paris, 299, ser. III,* 2, 35–38.

SAVIGNY, J.C., 1826 – In: Cuvier. Analyse des trav. acad. roy. sci. pendant l'année 1821, partie physique. *Mem. acad. roy. sci. inst. Fr.,* 5, 176–184.

Choice of Earthworm Species for Laboratory Tests

P.J. EDWARDS AND J.M. COULSON

ICI Agrochemicals, Environmental Sciences Department, Jealott's Hill Research Station, Bracknell, Berks, RG12 6EY, U.K.

Abstract

Comparisons in the susceptability (mortality) of up to 5 different species of earthworm were made in 42 comparisons of organic chemicals, mostly pesticides, incorporated into soil. There was no consistent relationship between species. However, *Apporectodea caliginosa* was frequently the most susceptable species, *Lumbricus terrestris* intermediate and *Eisenia fetida* least susceptable. In 39 comparisons differences in susceptability were within a factor of 10. *Eisenia fetida* is recommended for the initial screening of chemicals.

Very little comparative data are available for sub-lethal effects of different species.

The susceptability (mortality) of *Apporectodea caliginosa*, *Lumbricus terrestris* and *Eisenia fetida* to surface spray and granular formulations of insecticides, with and without grass cuttings on the soil surface was investigated. Differences in response between species when ›10 fold could be explained by earthworm behaviour.

Further comparative testing between species is required to 1) confirm an adjustment factor of 10 as adequate for E. fetida in the initial screening of chemicals and 2) identify the most appropriate species for sub-lethal testing.

Introduction

In the early 1980's the EEC and OECD required that industrial chemicals and pesticides be tested for toxicity to earthworms. The EEC brought together experts in the field of earthworm ecotoxicology to agree a test method. This method recommended that mature *Eisenia fetida* are exposed for 14 days to different concentrations of a chemical incorporated into an artificial soil. The method was examined by several laboratories in a 'Ring Test' (Edwards 1984).

There have been doubts about the sensitivity of *E. fetida*, a compost earthworm, and concern about it representing 'typical' earthworms. As a result of this other species have been promoted as test earthworms, especially *L. terrestris* (Dean-Ross 1982) and *A. caliginosa* (Martin 1985, 1986).

The risk of chemicals to earthworms has generally been assessed in a stepwise screening process such a that described by Kokta and Rothert (1992) and Greig-Smith (1992). The first step is a laboratory test designed to screen out chemicals safe to earthworms while other chemicals are tested in higher tier laboratory or field studies. The test organism must be well suited to laboratory use, preferably be easily cultured in the laboratory and provide a reliable prediction of the hazard to other species in the group of organisms it represents (Bouché 1992). At each tier of testing there should be a low risk of chemicals being falsely assessed as safe (false negative result). Species of earthworms do vary in their susceptability to pesticides. A reliable prediction of hazard may be acheived using species of low sensitivity if there is a consistent relationship between the test organism and the other species in the group it represents. In these circumstances the use of an 'adjustment factor' may be acceptable to align the results of the test organism to match the 'typical' or most sensitive species in the group.

Susceptability of an earthworm in soil is dependent on exposure and the toxicological response. Exposure may be dependent on the availability of the chemical, its distribution in the soil and earthworm behaviour, while the toxicological response will depend on toxicological activity and the ability of the earthworm to detoxify and excrete the chemical.

Any measure of toxicity to an organism depends on the experimental conditions and it is very difficult to compare the susceptability of earthworm species to chemicals where the experimental conditions vary. In this review an attempt has been made to look at the differences between species in their susceptabily under identical test conditions as much as possible. Differences observed have been explained where possible in terms of exposure and the toxicological response.

Comparative toxicity of pesticides to earthworms

The examination of laboratory toxicity for carbofuran to different species of earthworms is probably a good start for a critical review of the subject.

Gilman and Vardanis (1974) compared the toxicity of carbofuran to *E. fetida* and *Lumbricus terrestris* 5 days after injection and after 5 days exposure in soil. After injection the LC_{50} values for *E. fetida* and *L. terrestris* were 13ug/g worm and 2·4ug/g worm, respectively, a 6 fold difference. In soil the LC_{50} values for *E. fetida* and *L. terrestris* were 25 ppm and 13 ppm, only a 2 fold difference. *E. fetida* excreted 95% of an injected radioactive dose as carbofuran and the metabolite 3–hydroxy carbofuran in 2 days compared to only 10% in L. terrestris. Retained radioactivity was 50% unextractable and 50% unidentified metabolite in *L. terrestris*.

Stenersen (1979) compared the toxicity of 5 pesticides to 4 species of earthworms in the laboratory. In one test earthworms were dipped in pesticide solutions for 30 seconds then placed in clean soil (Table 1). In a

Table 1. The toxicity of pesticide solutions to earthworms after immersion (LC_{50} ug/ml) (Stenersen 1979).

Species	Aldicarb	Oxamyl	Carbaryl	Carbofuran	Gamma HCH
E. fetida	20–50	›200	200–800	400–800	400–800
A. caliginosa	3·1–6·3	›200	3·1–6·3	0·39–0·78	400–800
A. chlorotica	Not tested	›200	approx 200	5–10	400–800
L. rubellus	12·5–25	›200	25–50	5–10	400–800

Table 2. The toxicity of pesticides in soil to earthworms (LC_{50} mg/kg dry weight soil) (Stenersen 1979).

Species	Aldicarb	Carbaryl	Carbofuran	Paraoxon	Dithiocarbamate
E. fetida	16	›64	›64	32–64	16–32
A. caliginosa	4	‹4	‹4	32	16–32
A. chlorotica	4	4	4	‹32	8
L. rubellus	4–8	‹4	‹4	Not tested	16

second test earthworms were exposed to a range of pesticide concentrations in soil for 14 days (Table 2).

In both tests *E. fetida* showed substantially lower susceptability to carbofuran and carbaryl than the other species of earthworm. Susceptability between the other 3 chemicals tested by Stenersen (1979) in soil were similar. However, the difference in toxicity between *E. fetida* and *L. rubellus* for carbaryl was not consistent with the results of (Heimbach pers comm). In Heimbach's experiments the 28 day LC_{50} values were 27 and 32 mg/kg for *E. fetida* and *L. rubellus*, respectively in artificial soil test (OECD 207).

The differences in LC_{50} values between pesticides may be partially explained by the relative rate of uptake and elimination by earthworms. This has been demonstrated for carbofuran, oxamyl and parathion (Stenersen and Oien 1980, Stenersen 1979). This may also account for any differences between species with *E. fetida* more able to eliminate carbofuran than *L. terrestris* (Gilman and Vardanis 1974).

Kula (in prep) compared the toxicity of parathion and propoxur to five species of earthworm at different temperatures in artificial soil (Table 3). Like carbofuran there were large differences in susceptability for another carbamate, propoxur. Differences in species susceptability to parathion were relatively small.

Van Gestel and Ma (1990) examined the effect of different soils on the toxicity of 5 different chlorophenols, 2,4-dichloroanaline and 1,2,3-trichlorobenzene to *E. andrei* (*E. fetida* complex) and *L. rubellus*. There was

Table 3. The toxicity of parathion and propoxur in soil to earthworms (LC_{50} mg/kg dry weight soil) (Kula in prep)

Species	Temperature (°C)	Parathion	Propoxur
A. longa	15	60	0·8
A. caliginosa	15	116	1·4
A. chlorotica	15	206	10·8
L. terrestris	10–12	159	7·2
E. fetida	20	277	58

no appreciable difference in susceptability between species exposed to 2,4-dichloroanaline and 1,2,3-trichlorobenzene and on avereage *E. andrei* was 4 times more susceptible to all 5 chlorophenols than *L. rubellus*.

The author has compared the toxicity of 3 chemicals, benomyl, paraquat and chloracetamide evenly incorporated into a loamy sand soil to 3–5 species of earthworm after 28 days exposure (Table 4). In experiments with benomyl and paraquat, 4 *L. terrestris*, 10 *A. caliginosa*, 10 *E. fetida* were exposed together in 1kg soil in glass beakers with 3 replicates per concentration. In the experiment with chloroacetamide 3 *L. terrestris*, 5 *A. caliginosa*, 6 *E. fetida*, 3 *A. nocturna* and 2 *A. rosea* were exposed together in the same amount of soil and with the same numbers of replicates. Soil contained 2% organic matter, 8·8% clay, had a pH of 6·5, cation exchange capacity of 5·8 and was maintained at a moisture content of 20% (dry weight), equivalent to field capacity). The test was conducted at 15°C with a 16-hour light/8-hour dark photoperiod.

Table 4. The toxicity of benomyl, paraquat and chloracetamide in loamy sand soil to earthworms (LC_{50} mg/kg dry weight soil)

Species	Benomyl	Paraquat	Chloracetamide
L. terrestris	1–10	1000	54
A. caliginosa	1–10	580	40
A. nocturna	Not tested	Not tested	80
A. chlorotica	Not tested	Not tested	47
E. fetida	1–10	3200	75

Mortality data for species comparisons, under identical experimental conditions, show that *A. caliginosa* is frequently the most susceptable species to chemicals incorporated in soil followed by *L. terrestris* with *E. fetida* usually the least susceptable.

Further comparative data were compiled by Heimbach (1985), using data from Haque and Ebing (1983) and his own. The toxicity (LC_{50}) of 23 chemicals evenly incorporated in soil to *L. terrestris* and *E. fetida* were compared. *L. terrestris* were exposed in loamy sand soil with added peat (10%), at 10–15°C and 35% water holding capacity. *E. fetida* were exposed in an artificial soil (OECD 207) at 22°C and 35% moisture content. Their results are presented in Table 5. and show there is no consistent relationship between the relative susceptibility of *L. terrestris* and *E. fetida*.

To summarise, in 42 pesticide comparisons involving from 2 to 5 species, there was no consistent relationship between species. However, *A. caliginosa* was frequently the most susceptable species to chemicals incorporated in soil with *L. terrestris* of intermediate and *E. fetida* of least susceptability. With the exceptions of 3 comparisons, involving carbofuran, carbaryl and propoxur, differences in susceptability fell within a factor of 10. To put this difference in perspective, variability in LC_{50} values for *E. fetida* was about 10 fold between laboratories using the artificial soil test (OECD 207) (Edwards 1984) and 6 fold between 4 different soils in the laboratory (Van Gestel and Ma 1990).

Table 5. Comparison of LC_{50} values (mg/kg dry weight soil) for *E. fetida* and *L. terrestris* from Haque and Ebing (1983) (1) and Heimbach (1984) (2).

Chemical	*L.terrestris* (1)	*E. fetida* (1)	*E. fetida* (2)
Benomyl	3·5	27	19
Terbufos	4·6	6·6	Not tested
Carbofuran	4·7	28	5–10
Propoxur	5·2	Not tested	27
Methidathion	7·6	Not tested	4·8
Endosulfan	9·0	Not tested	9·4
Aldicarb	26	3·3	8
Fentinacetat+maneb	44	Not tested	28
Cu-oxichlorid	98	Not tested	900
Bupirimate	103	Not tested	401
Methamidophos	110	17	Not tested
Lindane	113	136	59
Calciumcyanamid	124	115	Not tested
Dialifos	174	Not tested	218
Triazophos	210	Not tested	116
Captan	237	Not tested	612
Atrazin	444	131	Not tested
Folpet	459	339	Not tested
Paraquat	›200	›200	Not tested
Triadimefon	›250	›250	Not tested
Chlormequat chloride	›460	›460	Not tested
Sodium chlorate	›750	›750	Not tested
Captafol	›800	Not tested	516

Sublethal effects such as growth and reproduction have been measured in the laboratory (Lofs-Holmin 1980, Martin 1986, Van Gestel *et al.* 1989, Kokta 1992). Lofs-Holmin has comparative data for *A. caliginosa* and *L. terrestris*. *A. caliginosa* was recommended as the most suitable species because the longer life-span and time to maturity of *L. terrestris* makes this species less suitable for use. The lack of comparative data for sublethal effects has made comparisons in species susceptability impossible.

Earthworm behaviour can effect their susceptability to chemicals through exposure. The amount of burrowing may effect exposure to soil incorporated chemicals (Martin 1982) and surface foraging to surface applications. This has been demonstrated in the laboratory (Curl 1987). *A. caliginosa* and *L. terrestris* were exposed for 8 weeks to C^{14} labelled cypermethrin incorporated into soil. Bioconcentration factors for *A. caliginosa* and *L. terrestris* were 30 and 8, respectively. The difference was consistent with their behaviour. *L. terrestris* has a permanent burrow and forages on the soil surface while *A. caliginosa* actively burrows through the soil when conditions are suitable. This difference in exposure has been confirmed in the field by Wheatley and Hardman (1968). Six species of earthworm from arable soil were analysed for organochlorine residues and concentrations were consistently higher in smaller and more shallow living species. The behaviour of *E. fetida* has more in common with *A. caliginosa* than *L. terrestris*, therefore *E. fetida* is probably a good behavioural model for chemicals incorporated in soil.

The way the chemical is applied to soil in a laboratory test and the presentation of potential food can both affect the apparent toxicity of a

Table 6. Effect of surface applied granules on earthworms in the laboratory.

Treatment	% Mortality		
	L. terrestris	*A. caliginosa*	*E. fetida*
Untreated Control	0	0	0
Untreated Control + grass	0	0	0
Granule A	78	56	67
Granule B	94	78	83
Granule C	0	0	0
Granule D	50	0	0
Carbofuran 0·2N	6	0	0
Carbofuran 0·2N + grass	6	44	50
Carbofuran 1N	44	22	33
Carbofuran 1N + grass	67	67	67
Carbofuran 5N	61	67	67
Carbofuran 5N + grass	44	44	33

Granules A–D were 4 different exploratory granules.
N – Normal rate (1·12 kgai/ha).

pesticide. The author has compared the effects of surface applied insecticide granules (Table 6) and the influence of grass cuttings placed over the granules on the soil surface to 3 species of earthworm. *E. fetida* (3), *L. terrestris* (3) and *A. caliginosa* (6) were exposed together in each of 3 replicates of 1 kg of loamy sand soil in glass beakers for 28 days. Soil moisture content was maintained at 20% (dry weight), at a temperature of 15°C and with a 12-hour light/12-hour dark photoperiod.

All 3 species were of similar susceptability to surface applied granules. The differences between *L. terrestris* and the other 2 species may be explained by exposure related to their foraging behaviour.

In another experiment the author exposed 4 species of earthworm to surface sprayed benomyl and chloracetamide with and without grass cuttings on the soil surface (Table 7). The soil surface and grass cuttings were sprayed by E jet (Even swath) moving down a track at a fixed speed to give a spray volume of 200l/ha. *E. fetida* (6), *L. terrestris* (3), *A. caliginosa* (6) and *A. nocturna* (3) were exposed together to a range of rates in each of 2 replicates. Glass beakers contained 1·3kg of a sandy loam soil (surface area about 130 cm^2). Earthworms were exposed for 28 days in soil at 23% moisture content (dry weight), a temperature of 15°C under lighting with a 16-hour light/8-hour dark photoperiod. Grass cuttings appear to have increased the exposure of *L. terrestris* to benomyl. This contrasts with *L. terrestris* being less exposed to chloracetamide on grass.

Table 7. Effect of surface sprayed benomyl and chloracetamide on earthworms in the laboratory.

Treatment	LC_{50} (mg/kg)		
	L. terrestris	*A. caliginosa*	*E. fetida*
Benomyl	3·6	3·5	5·6
Benomyl + grass	1·6	approx 20	>20
Chloracetamide	130	>320	400
Chloracetamide + grass	>320	>320	>320

It may be difficult to reproduce the normal behaviour of earthworms in the laboratory which is one of the main reason for testing effects on field populations. Bieri (1989) developed a funnel test for *L. terrestris* which was designed to allow their normal behaviour.

Discussion

Relatively few published studies compare the toxicity of different species of earthworms under identical experimental conditions. The results of this review conclude that there was no consistent relationship between species. However, *A. caliginosa* was frequently the most susceptable and *E. fetida* frequently the least susceptable when the chemical was incorporated in soil. In 39 out of 42 pesticide comparisons susceptability was within a factor of 10 for all species tested. A similar level of variability exists between laboratories and soils for a single species (Edwards 1984, Van Gestel and Ma 1990). Thus, the choice of species for laboratory tests assessing mortality and sublethal effects should be based on their suitability for laboratory experimentation and the ability to give consistent results.

There is considerable experience in the use of *E. fetida* in the laboratory, particularly concerning toxicity tests in artificial soil and rearing in cultures. It is recommended that *E. fetida* remains the standard laboratory species for initial laboratory screening ie OECD guideline 207. A factor of 10 will bring *E. fetida* in line with the most sensitive species with a false negative rate of about 7%. There is a virtual absence of comparative data for species and sub-lethal effects in laboratory studies.

For both lethal and sub-lethal tests there is a clear need for a programme of co-operative research to compare the susceptability of different species under identical conditions at different laboratories. In the case of lethal testing, to confirm the use of *E. fetida* as a test species in initial laboratory screening and 10 as the adjustment factor to lower false negative results to an acceptable level. In the case of sub-lethal tests (reproduction and growth) to establish the most appropriate species and set adjustment factors if neccessary.

References

BOUCHE', M.B. (1992). Earthworm species and ecotoxicological studies. Procedings of the International Working-Group on Earthworm Ecotoxicology.

CURL, E.A, EDWARDS, P.J., ELLIOTT, C. AND LEAHEY, J.P. (1987). The Conjugation and Accumulation of Metabolites of Cypermethrin by Earthworms. *Pesticide Science* **20**, 207–222.

DEAN-ROSS, D. (1982). Methods for the Assessment of the Toxicity of Environmental Chemicals to Earthworms. *Regulatory Toxicology and Pharmacology* **3**, 48–59.

EDWARDS, C.A. (1984). Report of the Second Stage in Development of a Standardized Laboratory Method for Assessing the Toxicity of Chemical Substances to Earthworms. *Report of the Commission of the European Communities EUR 9360 EN*.

GILMAN, A.P. AND VARDANIS, A. (1974). Carbofuran. Comparative Toxicity and Metabolism in the Worms Lumbricus terrestris and Eisenia foetida. *Journal Agr. Food Chem* **22**, 625–628.

GREIG-SMITH, P.W. (1992). Risk Assessment Approaches in the U.K. for the side-effects of pesticides on earthworms. Procedings of the International Working-Group on Earthworm Ecotoxicology.

HAQUE, A. AND EBING, W. (1983). Toxicity determination of pesticides to earthworms in the soil substrate. *Journal of Plant Diseases and Protection,* **90**, 395–408.

HEIMBACH, F. (1985). Comparison of laboratory methods, using Eisenia foetida and Lumbricus terrestris, for the assessment of the hazard of chemicals to earthworms. *Journal of Plant Diseases and Protection,* **92**, 186–193.

KOKTA, C. (1992). Measuring Effects of Chemicals in the Laboratory – Effect Criteria and Endpoints. Procedings of the International Working-Group on Earthworm Ecotoxicology.

LOFS-HOLMIN, A. (1980). Measuring Growth of Earthworms as a Method of Testing Sublethal Toxicity to Pesticides. *Swedish Journal of Agricultural Research,* **10**, 25–33.

MARTIN, N.A. (1982). The interaction between organic matter in soil and the burrowing activity of 3 species of earthworms (Oligochaeta: Lumbricidae). *Pedobiologia* **24:** 185–190.

MARTIN, N.A. (1985). Laboratory tests for toxicity of pesticides to earthworms. In Proc. 4th Australasian Conf on Grassl. Invert. Ecol. Lincoln College, Cantebury, 13–17 May, 1985. (R.B. Chapman, Ed) Caxton Press, New Zealand.

MARTIN, N.A. (1986). Toxicity of pesticides to Allolobophora caliginosa (Oligochaeta: Lumbricus). *New Zealand Journal of Agricultural Research,* **29**, 699–706.

STENERSEN, J. (1979). Action of Pesticides on Earthworms. Part I: The Toxicity of Cholinesterase-inhibiting Insecticides to Earthworms as Evaluated by Laboratory Tests. *Pesticide Science,* **10**, 66–74.

STENERSEN, J. AND OIEN, N. (1979). Action of Pesticides on Earthworms. Part IV: Uptake and Elimination of Oxamyl Compared with Carbofuran. *Pesticide Science* **11**, 396–400.

VAN GESTEL, C.A.M, VAN DIS, W.M., VAN BREEMEN, E.M. AND SPARENBOURG, P.M. (1989). Development of a Standardized Reproduction Toxicity Test with the Earthworm Species Eisenia fetida andrei using Copper, Pentachlorophenol and 2,4-Dichloroaniline. *Ecotoxicology and Environmental Safety,* **18**, 305–312.

VAN GESTEL, C.A.M. AND MA, W. (1990). An approach to quantitative structure-activity relationships (QSARs) in earthworm toxicity studies. *Chemosphere* **21**, 1023–1033.

WHEATLEY, G.A. AND HARDMAN, J.A. (1968). Organochlorine insecticide residues in earthworms from arable soils. *Sci. Fd Agric,* **19**, 221–225.

The Influence of Soil Characteristics on the Toxicity of Chemicals for Earthworms: A Review

C.A.M. VAN GESTEL

Department of Ecology and Ecotoxicology, Free University, De Boelelaan 1087, 1081 HV Amsterdam, The Netherlands

Abstract

From a review of the literature it appears that only a limited number of data are available on the influence of soil characteristics on earthworm toxicity.

The earthworm toxicity of organic chemicals was shown to be dominated by soils' organic matter content, and acute toxicity data can be extrapolated among soils on the basis of differences in organic matter content.

For ionizable organic chemicals, soil pH may also influence earthworm toxicity. For heavy metals the influence of soil characteristics is hard to quantify, as for each metal other soil properties seem to dominate bioavailability. Results reviewed indicate that for chemicals for which the influence of soil properties is hard to quantify, it may be possible to extrapolate earthworm toxicity data among soils on the basis of sorption data.

Results of acute toxicity studies indicate that this approach can also be used to extrapolate data obtained in the OECD artificial soil, which was specially developed for earthworm toxicity testing, to natural soils.

Introduction

Results obtained in laboratory toxicity tests should have some predictive value with respect to effects to be expected in field situations. Therefore both soil type and route of exposure in laboratory studies should resemble that of the natural situation. Preferably a substrate should be chosen which has the characteristics of natural soils with respect to behaviour and availability of the chemical. On the other hand it should be easy to sample and to standardize the test substrate (Committee to Review Methods for Ecotoxicology, National Research Council, 1981).

In studies on earthworm toxicity described in the literature many different soils, both natural and artificial, are used. For reasons of standardization the artificial soil substrate prescribed by OECD (1984) and EEC (1985) may be a good choice. This soil is made up by 10% peat, 20% kaolin clay and 70% sand; pH is adjusted to 6·0 ± 0·5 by adding some calcium carbonate. Edwards (1983) supported the choice of an artificial soil, by mentioning that its adsorptive capacity, expressed as its CEC (Cation Exchange Capacity), was similar to that of a typical loam soil. CEC however, is especially important for the sorption of cations (Bolt and Bruggenwert, 1976), and cannot tell much about the sorption of organic chemicals.

In this paper the literature on the influence of soil characteristics on earthworm toxicity will be reviewed. Attempts will be made to quantify the influence of soil properties on the earthworm toxicity of chemical substances.

Acute toxicity tests

HEAVY METALS

Jäggy and Streit (1982) reported that the acute toxicity of copper for *Octolasium cyaneum* in three different soils was significantly correlated with soils' organic matter content. From their results the following equation can be derived for the relationship between the 4–days LC_{50} and the organic matter content (%OM) of the test soils:

$$\log (LC_{50}) = 1{\cdot}476 + 1{\cdot}053 * \log (\%OM) \quad (n=3;\ r=0{\cdot}9997;\ p<0{\cdot}01)$$

From this equation it can be concluded that an increase of the soils' organic matter content by a factor of 2 will give rise to an increase of the LC_{50} by a factor of 2·075. Neither soil pH nor calcium content did exert a significant effect on copper toxicity.

Van Gestel and Van Dis (1988) studied the influence of soil pH and organic matter content on the acute toxicity of cadmium for *Eisenia andrei*, using three different soils: OECD (artificial soil: pH 7·0, 7·7% OM, 10·4% clay), Gilze+pH (pH 7·0, 1·7% OM, 4·3% clay) and Gilze (pH 4·1, 1·7% OM, 4·3% clay). The resulting 14 days LC_{50} values in these soils were >1000, >1000 (with 5% mortality) and 423 mg/kg, respectively, indicating that soil pH was the most important factor determining cadmium toxicity.

ORGANIC CHEMICALS

Van Gestel and Van Dis (1988) studied the influence of soil characteristics on the toxicity of three organic chemicals for *E. andrei* in the three soils mentioned before (OECD, Gilze+pH and Gilze). Their results are summarized in table 1.

From these results Van Gestel and Van Dis (1988) concluded that for chloroacetamide soil organic matter content does not influence toxicity, whereas soil pH does. For DCA soil organic matter content is determining toxicity, and pH has no influence. Toxicity of pentachlorophenol seems to be influenced by both soil pH and organic matter content; the pH effect can be explained on the basis of its dissociation constant pKa of 4.74.

Table 1. Acute toxicity of three chemicals for *Eisenia andrei* in three different soils (after Van Gestel and Van Dis, 1988). See text for soil properties.

Chemical	LC_{50} in mg/kg dry soil		
	OECD	Gilze+pH	Gilze
Chloroacetamide	40	42	13
3,4-dichloroaniline	240	134	134
pentachlorophenol	29	16	52

Van Gestel and Ma (1988, 1990) determined the influence of soil characteristics on the earthworm toxicity of 1,2,3-trichlorobenzene (TCB), 2,4-dichloroaniline (DCA) and five chlorophenols: 3-chlorophenol (MCP), 2,4-dichlorophenol (DCP), 2,4,5-trichlorophenol (TCP), 2,3,4,5-tetrachlorophenol (TeCP), and pentachlorophenol (PCP). They used the OECD artificial soil (pH 5·9; 8·1% OM) and three natural soils: KOBG (pH 4·8, 3·7% OM), HOLT (pH 5·6, 6·1% OM), and WAPV (pH 3·6, 15·6% OM). Their results (14 days LC_{50} values) are summarized in table 2.

Table 2. Acute toxicity of five chlorophenols, 2,4-dichloroaniline and 1,2,3-trichlorobenzene for *Eisenia andrei* and *Lumbricus rubellus* in four different soils (after Van Gestel and Ma, 1990). See text for abbreviations and soil properties.

Chemical	LC_{50} in mg/kg dry soil							
	Eisenia andrei				*Lumbricus rubellus*			
	KOBG	HOLT	OECD	WAPV	KOBG	HOLT	OECD	WAPV
MCP	75	134	130	423	150	342	247	633
DCP	134	240	172	423	352	486	322	680
TCP	46	76	63	164	235	316	362	875
TeCP	117	166	–	–	515	875	–	–
PCP	84	142	83	502	1206	1013	362	2298
DCA	142	285	319	824	201	304	190	580
TCB	134	240	133	547	115	200	195	563

Van Gestel *et al.* (1991) determined the acute toxicity of four chlorobenzenes for two earthworm species in two of the four soils mentioned in table 2. Their results are shown in table 3.

Van Gestel and Ma (1988, 1990) demonstrated that differences between soils almost completely disappeared when LC_{50} values in mg/kg dry soil were recalculated towards values in mg/l pore water using sorption coefficients. So, sorption data may be used to extrapolate earthworm toxicity data among

Table 3. Acute toxicity of four chlorobenzenes for *Eisenia andrei* and *Lumbricus rubellus* in two different soils (after Van Gestel *et al.*, 1991). See text for soil properties.

Chlorobenzene	LC_{50} in mg/kg dry soil			
	Eisenia andrei		*Lumbricus rubellus*	
	KOBG	OECD	KOBG	OECD
mono-	240	446	547	1107
1,4-di-	128	229	184	615
1,2,3,4-tetra	75	223	112	201
penta	134	238	115	201

soils. As sorption was shown to increase with increasing organic matter content of the soils, LC_{50} values can be extrapolated among soils just on the basis of differences in organic matter content. In case of ionizable compounds, like chlorophenols, this approach cannot be followed. For such chemicals, besides organic matter content, also soil pH influences sorption. Van Gestel *et al.* (1991) found that upon correction for dissociation of the data for chlorophenols, sorption was significantly correlated with lipophilicity of the test chemicals. This opens the way to extrapolation of acute earthworm toxicity data between soils using sorption data calculated by means of QSARs.

Heimbach and Edwards (1983) studied the influence of a number of variables on the toxicity of benomyl and chloroacetamide for *E. fetida* in artificial soil. Variation of the peat content between 2·5 and 10%, the use of kaolin or bentonite clays in amounts between 5 and 20%, and using pH values between 4·9 and 7·5 had little influence on the toxicity of these chemicals. Similarly, variation of soil moisture content (between 17·5 and 51%) did not influence toxicity to a great extent, and neither did temperature (between 10 and 26°C). In case of chloroacetamide, these results are not surprisingly, as this substance was chosen as reference substance in OECD and EEC test guidelines because it does not adsorb appreciably to (artificial) soil (Edwards, 1983). The fact that pH did not influence toxicity of chloroacetamide is contradictory to the findings of Van Gestel and Van Dis (1988) mentioned above. For benomyl, these results are rather unexpected, as for this compound sorption onto soils' organic matter and clay may influence earthworm toxicity (Lofs-Holmin, 1980).

Sublethal toxicity tests

HEAVY METALS

Bengtsson *et al.* (1986) determined the influence of soil pH on the toxicity of copper, cadmium and lead for *Dendrobaena rubida* in a sandy soil mixed with cattle dung (1:2) (final soil organic matter content: 7·7–11·7%). Extractability with 1 M ammonium acetate of lead and cadmium was significantly increased by decreasing soil pH: when pH was decreased by one unit extractability was 1·3–2 times higher. The extractability of copper was, however, not influenced by soil pH. Uptake of metals in the earthworms sometimes correlated with extractability, but at other times it did not. Accumulation of lead and cadmium in cocoons did not correlate with soil pH, but copper concentration of cocoons was twice as high at pH 6·5 compared to pH 5·5. Survival of adult worms was affected at low pH (4·5) for all three metals tested, and cocoon production was halved when pH was lowered from 6·5 to 4·5. For cadmium, copper and lead, at pH 6·5 cocoon production was stimulated by low metal concentrations. Even more striking was the effect of soil pH on the hatching rate: the number of hatching cocoons in the control decreased with decreasing soil pH, while upon exposure to the three metals at a concentration of 100 mg/kg the opposite occurred. As only two metal concentrations were tested in this study, no firm conclusions can be drawn on the relation between soil pH and earthworm toxicity of the metals.

Van Gestel and Hoogerwerf (in prep.) determined the influence of soil pH on the toxicity of aluminium for *E. andrei* in artificial soil. Effects on growth and reproduction were studied according to a method described by Van Gestel *et al.* (1989); worms were exposed for 6 weeks (two periods of 3 weeks). Results in terms of NOEC values (No-Observed-Effect-Concentrations) are summarized in table 4.

Table 4. Influence of soil pH on the effect of aluminium on *Eisenia andrei* in artificial soil (6 weeks exposure).

Parameter	NOEC (mg Al/kg dry soil) at pH		
	3·4	4·3	7·3
survival	320	≥1000	≥1000
growth	100	≥1000	32#
cocoon production	100	100	100
cocoon fertility	100	≥1000	≥1000
juveniles/fertile cocoon	100*	≥1000	≥1000

* due to a very low number of cocoons reliability of this value is low
# at higher concentrations growth was significantly increased

From table 4 it can be concluded that low soil pH significantly increased aluminium toxicity. At the highest pH tested, earthworm growth was significantly increased (not dose-related) at high aluminium concentrations in soil. The effect of aluminium on cocoon production seems not to be influenced by soil pH. At pH 3·4, all worms died at 1000 mg Al/kg. At this lowest pH level cocoon production was almost completely inhibited at 320 mg Al/kg, whereas at pH 4·3 and 7·3 it was only halved at this concentration. It should further be noted that in the control groups cocoon production was significantly reduced at pH 3·4 compared to the two higher pH soils. The extractability with 1 N calcium chloride of aluminium appeared to decrease with increasing soil pH. The effects on growth and cocoon production could, however, only partially be related to the amount of free aluminium in soil. Apparently, other factors also played a role.

Ma (1988) determined the influence of copper on the reproduction of *Aporrectodea caliginosa* and *Lumbricus rubellus* in a sandy soil (4–6% OM, 2–4% clay) in relation to soil pH. In untreated soil *A. caliginosa* appeared to be more sensitive for pH changes than *L. rubellus*. Cocoon production by *L. rubellus* was not affected by soil pH (values between 3·9 and 6·1); cocoon production by *A. caliginosa* was optimal at a soil pH of c. 4·9 and more than 50% reduced at pH 3·9. From the paper of Ma (1988), EC_{50} values for the effect of copper on cocoon production in relation to soil pH were estimated. For this purpose cocoon production was estimated from figure 1 in the paper of Ma (1988), and EC_{50} values were calculated using a logit model. The results of this approach are summarized in table 5.

From table 5 it can be concluded that the effect of soil pH on copper toxicity is different for the two species tested. For *A. caliginosa* soil pH does not affect copper toxicity significantly, whereas for *L. rubellus* EC_{50} values

Table 5. EC_{50} values for the influence of soil pH on the effect of copper on the cocoon production by *Aporrectodea caliginosa* and *Lumbricus rubellus* in a sandy soil (based on data estimated from Ma, 1988).

Species	EC_{50} (in mg Cu/kg dry soil) at pH			
	3·9	4·2	4·9	5·6–6·1
A. caliginosa	99	96	99	112
L. rubellus	77	105	150	177

for the effect of copper on cocoon production decrease by a factor of about 2 when soil pH is decreased by 2 units.

Ma (1984) studied the effect of copper on growth, cocoon production and litter degradation by *L. rubellus* in two different soils: a loamy sand (pH 4·8, 5·7% OM, 2% clay) and a sandy loam (pH 7·3, 3·4% OM, 17% clay). Cocoon production was more severely reduced in the sandy loam, resulting in NOEC values of 130 and 65 mg Cu/kg in loamy sand and sandy loam, respectively. This difference is probably induced by the lower organic matter content of the latter soil. In the sandy soil the influence of soil pH was studied. Cocoon production was dose related decreased at pH 4·8 and 6·0, but at pH 7·0 the highest dose tested (278 mg Cu/kg dry soil) did not affect cocoon production.

Ma (1983) noted that there was a great difference in the performance of *L. rubellus* between the two soils: in the sandy loam worm growth, cocoon production, cocoon hatchability and litter decomposition were much higher than in the loamy sand. Considering the differences in the behaviour of *L. rubellus* between the two soils tested, Ma (1984) concludes: 'It is apparent that the factors determining copper toxicity may be complex, and in addition are not necessarily limited to abiotic soil conditions determining the physico-chemical copper activity in soil.'

ORGANIC CHEMICALS

Ma (1983) determined the effect of lindane on growth, reproduction and litter decomposition of *L. rubellus* in the same two soil types. In sandy loam, lindane appeared to be more toxic, with NOEC values of <12 and 12–$\geq$30 mg/kg for sandy loam and loamy sand, respectively. Like for copper, this difference may be attributed to the lower organic matter content of the first soil.

Lofs-Holmin (1982) determined the influence of benomyl on the reproduction of *A. caliginosa* in mixtures of sandy (1·2% OM, 4% clay) or clayey (4·6% OM, 53% clay) soils with farmyard manure (1:1). It appeared that the sandy soil was unsuitable for such tests with this earthworm species, as already in the control some mortality occurred and no cocoons were produced. In the sandy soil benomyl appeared to be much more toxic than in the loamy soil, resulting in a much higher dose-related mortality in the first soil. On one hand this can be attributed to the lower adsorption capacity of the sandy soil. On the other hand sensitivity may be increased by the less optimal conditions in the sandy soil.

In a study on the growth and sexual development of *A. caliginosa, Allolobophora chlorotica, A. rosea* and *L. terrestris*, using similar substrates of sand or clay with farmyard manure, Lofs-Holmin (1980) found similar effects. Growth of the latter three species was better in the clayey soil. Toxicity of benomyl and TCA was much higher in the sandy soil. Lofs-Holmin (1980) concludes that the adsorptive capacity of the farmyard manure probably is low, as characteristics of the soils (especially difference in clay content) dominate the effect.

Van Gestel *et al.* (1992a) found a negative correlation between growth and cocoon production of *E. andrei* in artificial soil. This phenomenon could only partially be explained on the basis of variation in soil characteristics. At soil pH values exceeding 7·0 cocoon production was significantly reduced; pH 5·0–6·0 appeared to be optimal. As a result of this phenomenon, in reproduction toxicity tests either growth or reproduction may be most sensitive. Van Gestel *et al.* (1992b) therefore recommend to include both parameters as endpoints in an earthworm reproduction toxicity test.

Discussion

From the literature cited in this paper, it can be concluded that only a limited number of data is available on the influence of soil characteristics on earthworm toxicity. And from the data available, it is hard to quantify the influence of certain soil properties on the effect parameters studied. Furthermore, it is suggested by some authors that differences between soils may induce differences in earthworm behaviour (Ma, 1983, 1984, 1988; Lofs-Holmin, 1980, 1982; Van Gestel *et al.*, 1992a). As this was only reported for sublethal toxicity tests, especially the sensitivity of sublethal parameters, like growth and reproduction, may be affected by this.

HEAVY METALS

For metals, the situation is quite complicated. The results presented by Bengtsson *et al.* (1986) and Jäggy and Streit (1982) demonstrate that for the toxicity of copper for *O. cyaneum* and *D. rubida* soil pH is not so important. Other soil properties, like organic matter content, seem to have greater influence on the bioavailability of copper. When soil pH is taken as the only variable, it may influence earthworm toxicity of copper, as is demonstrated by Ma (1984, 1988). Results derived from Ma (1988) demonstrate, however, that the effect of soil pH on copper toxicity is also dependent on the earthworm species tested. For the effect of cadmium and lead on *D. rubida* soil pH is much more important, as is indicated by the results of Bengtsson *et al.* (1986). Soil pH does not affect the influence of aluminium on cocoon production by *E. andrei*, but it does affect the effect on survival, reproduction and growth (Van Gestel and Hoogerwerf, in prep.).

A toxic effect of a chemical on an organism generally is the resultant of uptake resulting in a critical concentration in its tissues. Bioavailability can therefore also be determined as the degree of accumulation in earthworms.

So, for heavy metals reference can be made to the literature on this aspect to get more insight into the soil variables determining bioavailability. Ma (1982) concluded that BCF values for the uptake of cadmium, lead and zinc from soil by earthworms increased by a factor of c. 5 when soil pH decreased from 6·5 to 4·5. For these three metals soil pH and CEC were the most important factors determining uptake by earthworms. For copper, soil pH was not important, and earthworm concentrations were more related with soils' organic matter content and CEC. According to Ma *et al.* (1983) also soil organic matter influenced the uptake of lead from soil by earthworms. Morgan and Morgan (1988) found a significant relationship between the metal content of earthworms and soil concentrations; for copper and zinc this relationship could not be improved by including soil pH nor CEC, whereas for cadmium soil organic matter content significantly improved the relationship and so did soil pH and CEC for lead. Beyer *et al.* (1987) concluded from an analysis of the relationship between earthworm metal concentrations and soil variables, that 'measuring soil variables did not substantially improve the ability to predict earthworm concentrations'. The only exception was soil pH, which was associated with usually high or low concentrations of the metals studied.

From all these data it has to be concluded that in case of heavy metals, it is hard to quantify the influence of soil characteristics on the bioavailability and the toxicity for earthworms.

ORGANIC CHEMICALS

From the literature on acute toxicity, it may be concluded that for non-ionizable organic chemicals earthworm toxicity data can be extrapolated between soils on the basis of differences in organic matter content. This was shown for acute toxicity data on chlorobenzenes and dichloroaniline by Van Gestel and Ma (1990) and Van Gestel *et al.* (1991). In case of ionizable chemicals, like chlorophenols, soil pH may significantly affect toxicity. In such cases correction of the toxicity data for dissociation may improve the extrapolation among soils (Van Gestel *et al.*, 1991). On the other hand, for such chemicals sorption coefficients can be used to extrapolate toxicity data between soils (Van Gestel and Ma, 1990; Van Gestel *et al.*, 1991).

THE USE OF SORPTION DATA

As mentioned above, results presented by Van Gestel and Ma (1990) and Van Gestel *et al.* (1991) demonstrate that for organic chemicals sorption data may be useful in extrapolating earthworm toxicity data from one soil to another. This approach may also be useful in case of heavy metals. No literature data was found, however, to support this assumption. But it may be referred to the equilibrium partitioning concept developed in aquatic ecotoxicology for determining the ecotoxicological risk of pollutants in sediments. For this purpose it is assumed that organisms are mainly exposed via the sediment pore water, and that pore water concentrations can be predicted on the basis of sorption data (Ziegenfuss *et al.*, 1986; Shea, 1988; Giesy and Hoke, 1990). This approach also enables the extrapolation of toxicity data among sediments on the basis of differences in sorption behaviour.

Two remarks should be made with respect to the use of sorption data. Firstly, sorption of chemicals onto soils can be affected by soil moisture content. In a theoretical way, Boesten (1990) demonstrated that for chemicals with a Kd › 1 the water/soil ratio does no longer influence sorption. So for such chemicals soil moisture content does not influence pore water concentrations. Briggs (1990) concluded that chemicals with an octanol/water partition coefficient (log Kow) ›2 will have a Kd ›1 in soils with 1·5% organic carbon (equal to c. 2·6% OM). It can be concluded that for such chemicals in soils with ›2·6% organic matter earthworm toxicity will be independent of soil moisture content. As a consequence, for chemicals with log Kow ‹2 sorption coefficients cannot be used to extrapolate toxicity data between soils without taking into account differences in soil moisture content.

A second remark to be made with regard to the use of sorption data, is that the equilibrium partitioning method mentioned above may no longer be valid for highly lipophilic chemicals. For chemicals with log Kow ›5.0 oral uptake may become important, and exposure via pore water may no longer be the most important route of uptake (Landrum and Robbins, 1990).

THE USEFULNESS OF OECD ARTIFICIAL SOIL

The results presented by Van Gestel and Ma (1990) and Van Gestel *et al.* (1991) support the conclusion that the artificial soil prescribed by OECD (1984) and EEC (1985) is a good choice for earthworm toxicity tests on organic chemicals. It fulfils the requirements with respect to representativity for both behaviour and bioavailability of the chemicals. Results obtained in acute toxicity tests in artificial soil can be extrapolated to natural soils. The artificial soil can easily be standardized. To further improve standardization, it should be noticed that differences between lots of peat and or kaolin clay used may induce differences in moisture retention and pH of the artificial soil. The way the peat is sieved or ground may affect sorption of chemicals to the artificial soil. These factors need further standardization before a completely standardized artificial soil can be obtained. Finally, it should be noted that the pH of the artificial soil may increase by about 0·5 units during a 2 week toxicity test (Van Gestel and Van Dis, 1988); this is of importance for chemicals which behaviour in soil is pH dependent (e.g. metals and chlorophenols).

No data is available to enable comparison of sublethal effects of chemicals in artificial soil and in natural soils. Further research on this aspect is urgently needed.

References

BENGTSSON, G., GUNNARSSON, T., RUNDGREN, S. (1986) Effects of metal pollution on the earthworm *Dendrobaena rubida* (Sav.) in acidified soils. *Water, Air and Soil Pollution* **28,** 361–383.

BEYER, W.N., HENSLER, G, MOORE, J. (1987) Relation of pH and other soil variables to concentrations of Pb, Cu, Zn, Cd, and Se in earthworms. *Pedobiologia* **30,** 167–172.

BOESTEN, J.J.T.I. (1990) Influence of solid/liquid ratio on the experimental error of sorption coefficients in pesticide/soil systems. *Pesticide Science* **30,** 31–41.

BOLT, G.H., BRUGGENWERT, M.G.M. (1976) Soil chemistry. *A. Basic elements.* Elsevier, Amsterdam.

BRIGGS, G.G. (1990) Predicting the behaviour of pesticides in soil from their physical and chemical properties. *Phil. Trans. R. Soc. Lond. B* **329,** 375–382.

Committee to Review Methods of Ecotoxicology. National Research Council (1981) *Testing for effects of chemicals on ecosystems.* National Academic Press, Washington D.C.

EDWARDS, C.A. (1983) *Report of the second stage in development of a standardized laboratory method for assessing the toxicity of chemical substances to earthworms.* Commission of the European Communities, DOC XI/83/700, Brussels.

EEC (1985) *EEC Directive 79/831. Annex V. Part C: Methods for the determination of ecotoxicity. Level I. C (L1)4: Toxicity for earthworms. Artificial soil test. DG XI/128/82. Final.*

GIESY, J.P., HOKE, R.A. (1990) Freshwater sediment quality criteria: toxicity bioassessment. In: Sediments: Chemistry and Toxicity of In-place Pollutants. Baudo, R., Giesy, J., Muntau, H. (Eds.). Lewis Publishers Inc., Ann Arbor. pp. 265–348.

HEIMBACH, F., EDWARDS, P.J. (1983) The toxicity of 2-chloroacetamide and benomyl to earthworms under various test conditions in an artificial soil test. *Pesticide Science* **14,** 635.

JÄGGY, A., STREIT, B. (1982) Toxic effects of soluble copper on *Octolasium cyaneum* Sav. (Lumbricidae). *Revue Suisse Zoologie* **89,** 881–889.

LANDRUM, P.F., ROBBINS, J.A. (1990) Bioavailability: how sediment-associated contaminants are accumulated by benthos or is interstitial water the answer. In: Sediments: Chemistry and Toxicity of In-place Pollutants. Baudo, R., Giesy, J., Muntau, H. (Eds.). Lewis Publishers Inc., Ann Arbor. pp. 237–263.

LOFS-HOLMIN, A. (1980) Measuring growth of earthworms as a method of testing sublethal toxicity of pesticides. *Swedish Journal of agricultural Research* **10,** 25–33.

LOFS-HOLMIN, A. (1982) Measuring cocoon production of the earthworm *Allolobophora caliginosa* (Sav.) as a method of testing sublethal toxicity of pesticides. *Swedish Journal of agricultural Research* **12,** 117–119.

MA, WEI-CHUN (1982) The influence of soil properties and worm-related factors on the concentration of heavy metals in earthworms. *Pedobiologia* **24,** 109–119.

MA, WEI-CHUN (1983) Regenwormen als bio-indicators van bodemverontreiniging. *Bodembescherming* **15**. Staatsuitgeverij, The Hague, the Netherlands.

MA, WEI-CHUN (1984) Sublethal toxic effects of copper on growth, reproduction and litter breakdown activity in the earthworm *Lumbricus rubellus*, with observations on the influence of temperature and soil pH. *Environmental Pollution (Series A)* **33,** 207–219.

MA, WEI-CHUN (1988) Toxicity of copper to lumbricid earthworms in sandy agricultural soils amended with Cu-enriched organic waste materials. *Ecological Bulletins* **39,** 53–56.

MA, W., EDELMAN, TH., VAN BEERSUM, I., JANS, TH. (1983) Uptake of cadmium, zinc, lead, and copper by earthworms near a zinc-smelting complex: influence of soil pH and organic matter content. *Bulletin of Environmental Contamination and Toxicology* **30,** 424–427.

MORGAN, J.E., MORGAN, A.J. (1988) Earthworms as biological monitors of cadmium, copper, lead and zinc in metalliferous soils. *Environmental Pollution* **54,** 123–138.

OECD (1984) *Guideline for testing of chemicals no. 207. Earthworm, acute toxicity tests.* Adopted 4 April 1984.

SHEA, D. (1988) Developing national sediment quality criteria. Equilibrium partitioning as a means of evaluating sediment quality. *Environmental Science and Technology* **22,** 1256–1261.

Van Gestel, C.A.M., Van Dis, W.A. (1988) The influence of soil characteristics on the toxicity of four chemicals to the earthworm *Eisenia fetida andrei* (Oligochaeta). *Biology and Fertility of Soils* **6,** 262–265.

VAN GESTEL, C.A.M., VAN DIS, W.A., VAN BREEMEN, E.M., SPARENBURG, P.M. (1989) Development of a standardized reproduction toxicity test with the earthworm species *Eisenia fetida andrei* using copper, pentachlorophenol, and 2,4-dichloroaniline. *Ecotoxicology and Environmental Safety* **18,** 305–312.

VAN GESTEL, C.A.M., MA, WEI-CHUN (1988) Toxicity and bioaccumulation of chlorophenols in earthworms, in relation to bioavailability in soil. *Ecotoxicology and Environmental Safety* **15,** 289–297.

VAN GESTEL, C.A.M., MA, WEI-CHUN (1990) An approach to quantitative structure-activity relationships (QSARs) in earthworm toxicity studies. *Chemosphere* **21,** 1023–1033.

VAN GESTEL, C.A.M., MA, WEI-CHUN, SMIT, C.E. (1991) Development of QSAR's in terrestrial ecotoxicology: earthworm toxicity and soil sorption of chlorophenols, chlorobenzenes and dichloroaniline. *Science of the Total Environment*, **109/110**, 589–604.

VAN GESTEL, C.A.M., DIRVEN-VAN BREEMEN, E.M., BAERSELMAN, R. (1992a) Influence of environmental conditions on the growth and reproduction of the earthworm *Eisenia andrei* in an artificial soil substrate. *Pedobiologia* (in press).

VAN GESTEL, C.A.M., DIRVEN-VAN BREEMEN, E.M., BAERSELMAN, R., EMANS, H.J.B., JANSSEN, J.A.M., POSTUMA, R., VAN VLIET, P.J.M. (1992b) Comparison of sublethal and lethal criteria for nine different chemicals in standardized toxicity tests using the earthworm *Eisenia andrei. Ecotoxicology and Environmental Safety* (in press).

ZIEGENFUSS, P.S., RENAUDETTE, W.J., ADAMS, W.J. (1986) Methodology for assessing the acute toxicity of chemicals sorbed to sediments: testing the equilibrium partitioning theory. In: *Aquatic toxicology and environmental fate: Ninth volume*. POSTON, T.M., PURDY R. (Eds.), *ASTM STP* **921.** American Society for Testing and Materials, Philadelphia.

Measuring Effects of Chemicals in the Laboratory: Effect Criteria and Endpoints

C. KOKTA

Federal Biological Research Centre for Agriculture and Forestry, Department for Plant Protection Products and Application Techniques, Zoology Division, Messeweg 11/12, D-3300 Braunschweig, Germany

Abstract

Laboratory tests should give ecologically relevant information on effects of chemicals to earthworms. Standardization of the test design for laboratory tests is required. Mortality, body weight change and reproduction are discussed as the main endpoints for assessment of adverse effects of chemicals to earthworms in the laboratory.

Introduction

Investigation of adverse effects of chemicals on non-target organisms should start in the laboratory. Laboratory tests give information on the toxicity of chemicals under standardized conditions. Risk assessment and if necessary further testing must follow to get information on possible adverse effects under field conditions.

Keeping in mind a stepwise testing procedure it has to be specified which criteria a laboratory test must fulfill and what kind of result is required. The following criteria seem to be essential:

- ecological significance of the effects measured in the laboratory test
- validation of the results found in the laboratory by correlating them with effects observed in the field
- standardization of the test system
- reproducibility
- effects measurable as simple as possible to improve practicability and handling and thereby reduce possible errors
- proper statistical treatment of data

In laboratory testing of earthworms the following endpoints are used or can be used to get information on possible adverse effects:

- mortality
- body weight change of adult earthworms
- reproduction
- juvenile development
- behaviour
- morphological effects
- physiological effects
- decomposition

In the following these main endpoints, some of them being established in existing international guidelines (e.g. OECD, 1984; EEC, 1985) will be discussed concerning their ecological significance, practical use and possibilities of standardization.

Mortality

As Heimbach (1991, in press) pointed out there is a good correlation between laboratory data on mortality (LC_{50}) assessed with *Eisenia fetida* and effects occuring in the field.

Mortality is established as main endpoint in acute toxicity tests with *Eisenia fetida* (e.g. OECD, 1984; EEC, 1985). For the purpose of standardization, an artificial substrate ('artificial soil') has been developed by Edwards (1983).

Mortality may also serve as a main endpoint when investigating effects on species of arable land in the laboratory. For means of standardization however it is necessary to rear these species in the laboratory. Only by laboratory breeding it is guaranteed that animals of the same age and with known life history are used. Although there are methods for laboratory breeding of species of arable land (Lofs-Holmin, 1982; Martin, 1986) it is still difficult to rear them in sufficient numbers.

Different sensitivity of *Eisenia fetida* and other species occurs (e.g. Haque & Ebing, 1983, Heimbach, 1985) and has to be taken into account by an uncertainty factor when interpreting the results.

Adult animals are generally used for determining mortality. Juvenile earthworms are more difficult to handle and there seem to be no substantial differences concerning sensitivity of adults and juveniles (Kula, pers. comm.).

Usually, mortality is determined by means of the LC_{50}, i.e. the concentration which is lethal for 50% of the test animals. The slope of the dose-response graph gives valuable information.

When the guideline for testing *Eisenia fetida* was developed, tests were run over 28 days to determine mortality, body weight development and reproduction. Since mortality was chosen as the main endpoint and since there was not much additional mortality between day 14 and day 28, a test duration of 14 days was fixed. To get some information on the time-effect relationship, counting of dead earthworms after 7 days of exposure was introduced so that there are now data on 7 day- and 14 day-mortality within the test.

The question of reproducibility of mortality can be answered from two international ringtests organised by C.A. Edwards (1983, 1984) in the course of the development of the test on acute toxicity. The result was that the artificial soil test can provide reproducible results on toxicity with an accuracy within an order of magnitude for most of the tested chemicals.

Additional data are available from a small ring test by members of a german working group on earthworm testing. In the german tests the LC_{50} values of 4 pesticides varied by a factor of 3 to 13 (range from minimum to maximum value). This variation has to be considered by introducing an uncertainty factor when using the results for regulatory decisions.

So far, a chemical concentration of 1000 mg active ingredient/kg dry substrate has been set as an upper limit for testing. Chemicals with an LC_{50} of more than 1000 mg active ingredient/kg dry substrate are believed to cause no problems for earthworms in the field. In the case of pesticides, the formulated product has to be tested and an LC_{50} above 1000 mg of the formulated product is considered to be without harm to earthworms. With pesticides, the highest recommended application rate is very important for risk assessment and so it might be useful to set a relevant concentration as an upper limit, e.g. the 100-fold EEC (EEC = Estimated Environmental Concentration). For calculation of EEC the following values are used as a first approach: a mean bulk density of 1·5 g/cm^3 and an even distribution of the chemical in the upper 2·5 cm of the soil.

When the range-finding-test gives the information that mortality will occur only at concentrations above 1000 mg formulated product/kg dry substrate, the question arises whether a limit test is sufficient to verify the results of the range finding test. Concerning mortality this procedure will be sufficient but a certain variablity of results must be taken into account and other test criteria like body weight might demand additional concentrations to be tested.

At the end of a test, animals often are alive but obviously have suffered irreparable damage. Perhaps these animals would die if the tests were prolonged. Therefore it must be stressed that the physical condition of the earthworms has to be described carefully.

Body weight change of adult earthworms

Change of individual body weight of adult earthworms during the test is an endpoint which can be measured easily within the test on acute toxicity.

When trying to get information on body weight change within the test on acute toxicity, some problems have to be considered. As the earthworms are not fed in this test, all animals including those of the control boxes might loose body weight up to a certain degree. It is difficult then to differentiate between effects of the chemical and starvation effects. From 72 trials with pesticides from different laboratories (unpublished), 18% showed more than 10% loss of body weight in the control and 8% more than 15% loss of body weight. To minimize the risk of wrong interpretation a quality criterion should be introduced by limiting the tolerated weight loss in the control. But it is hardly possible to prevent weight loss in a test without food. The only possibility is to control the water content of the test boxes carefully as weight loss of animals can be caused by desiccation of the substrate. Also in trials without food the variability of weight change may be higher than in trials with food.

According to the OECD-guideline body weight development and mortality are determined in the same test. This can cause some problems. Often, especially at high concentrations, there are 2 or 3 remaining earthworms that have gained weight whereas the others are already dead. It should be defined very clearly whether determination of individual life weight of surviving earthworms or total biomass of a test container is wanted. To solve this problem, data on body weight development should be used only when at most 20% mortality occurs.

Concerning the determination of sublethal effects on species of arable land in the laboratory, body weight change might be the only suitable endpoint because reproduction assessment takes a long time and might show a great variability. Also, such a test might be limited by substrate exhaustion. Ma (1983) found that body weight is a suitable parameter for *Lumbricus rubellus*, but larger amounts of soil are necessary.

Reproduction

Reproduction success is a very important factor for population dynamics. The use of *Eisenia fetida* in reproduction tests has several advantages compared to other species. First, *Eisenia fetida* has a short generation time. Second, there are several hatchlings per cocoon (Hartenstein *et al.*, 1979) so that the determination of reproduction parameters is possible within an appropriate time. Several parts of the reproduction process can be investigated separately:

- cocoon production
- hatching rate
- survival of juveniles

To collect information on these processes separately, it is necessary to destroy the test system. For example the determination of the cocoon production needs washing and sieving of the test substrate while the determination of hatching rate needs a method to extract hatchlings from the substrate.

Body weight change and reproduction

Body weight change will give information on sublethal effects, but as Neuhauser & Callahan (1990) pointed out it might not always be the most sensitive parameter compared for example to reproduction.

In the first international ringtest on earthworm toxicity measurement of cocoon production was dropped because standardization was difficult and weight change yielded enough information on sublethal effects. Another effort to introduce sublethal parameters, especially reproduction, was done by a working group in Germany in 1989. This group tried to determine effects on mortality, biomass and reproduction with and without food in the OECD-guideline.

Concerning fertility parameters, there was no overall difference in sensitivity between cocoon production and hatching rate. In 6 trials out of 8 hatching rate and cocoon production showed the same sensitivity and in 2 trials hatching rate was more sensitive by 1 concentration step than cocoon production. In 10 out of 20 trials the sensitivity of weight change and cocoon production was the same. In 6 trials cocoon production was more sensitive than body weight and in 4 trials the opposite was true. All differences in sensitivity comprised 1 concentration step, only in one case it comprised two steps.

Heimbach (unpublished results) found that 21 from 30 trials revealed the same sensitivity concerning body weight and cocoon production. Cocoon production was more sensitive than weight change 8 times and one time the opposite occurred.

Van Gestel (submitted) found a negative correlation between reproduction and growth of adult earthworms. Therefore it is necessary to look at both reproduction and weight change together. Own research also led to the conclusion that animals which gain much weight during the test produce less juveniles per adult. This correlation may be obscured by other factors, but it may lead to the conclusion that animals used in a reproduction test should not be in a growing phase any longer so that a certain fertility is granted. Perhaps the initial weight range of the animals should be defined more closely than it is described in the OECD-guideline with the range of 300 to 600 mg. Another possibility might be to ensure that only reproducing animals are introduced into the test. This might be done, for example, by a prolonged preincubation period.

From several trials concerning the variability of body weight change and reproduction success within one laboratory, it can be concluded that due to the complex relations of body weight change and reproduction these should be investigated together.

From the ecological point of view reproduction measurements in the laboratory should comprise the whole process. Determination of the number of juvenile earthworms after a certain time of exposure of adults and juveniles together are believed to give the most valuable infomation. Two methods concerning a reproduction toxicity test on *Eisenia fetida* are under development: the approaches of Van Gestel *et al.* (1988, 1989) are based on the OECD-guideline and examine cocoon production in contaminated substrate and hatchability in uncontaminated substrate while in the second method by Kula & Kokta (1991, in press) the cocoons remain in the undisturbed contaminated soil and the number of juveniles after 6 to 10 weeks of exposure to the chemical is determined. The second method seems to be closer to outdoor conditions and perhaps should therefore be preferred for testing pesticides. Further research is needed on a comparison of these methods.

The question whether determination of body weight change is sufficient for predicting sublethal effects on earthworm populations is still under discussion.

Juvenile development

Besides acute toxicity, sublethal effects are of great importance for population development. In addition to weight change of adult earthworms it is possible to look for growth and sexual development of juvenile earthworms. But as Van Gestel (unpublished) found, sensitivity of growth and sexual development of juveniles is similar to growth and reproduction of adults. Additionally, handling of juveniles is difficult.

Tests on growth rate (Martin, 1986; Pizl, 1988) may be useful for getting detailed information in special cases.

Behaviour

Earthworm behaviour is an important endpoint although it is difficult to describe and express in numbers.

When food is provided feeding activity can give information on possible repellent or immobilisation effects by chemicals. Bieri *et al.* (1989) developed a test which allows observation of the behaviour of single individuals of *Lumbricus terrestris*. As the soil surface of the test boxes is very small it is hardly possible to get information on relevant concentrations of certain pesticides. Feeding activity of earthworms in the laboratory can also be examined by castings, especially with *Lumbricus terrestris*, but it is difficult to use this parameter quantitatively.

A repellent effect of a short living pesticide probably does not impair earthworms very much whereas a long living pesticide with a repellent effect might result in a lowered population density of deep-burrowing species because food for the deep-burrowing species is available on the soil surface only.

Increased activity may also occur as a response to a chemical. In tests according to OECD-guideline No. 207 it is sometimes observed that animals try to avoid contact with the contaminated soil. Increased activity is of ecological importance as it can lead to population movements in the field and to death by sunlight or increased predation.

Morphological effects

Morphological changes which can be observed directly in test organisms are for example protrudings or constrictions on the body surface. Sometimes earthworms loose parts of their body. If morphological changes occur they should be judged very critical. Stenersen (1979) found in some pesticides that the first segments still reacted to a stimulus whereas most of the body was rotten.

Physiological effects

A first hint of stress on earthworms which can easily be examined is an increased mucus production. Sometimes enzyme activities are reported to be suitable for measuring chemical stress, for example cholinesterase activity

(Stringer & Wright, 1976). Neurotoxic effects of pesticides such as change of nerve fiber conduction velocities were reported by Drewes & Vining (1984) and Drewes *et al.* (1987). Often these endpoints need an expensive laboratory equipment. Probably the use will therefore be restricted to special questions. There is a need of investigation on the importance of these effects on population dynamics in the field.

Decomposition

Earthworms play an important role in decomposition processes. Trying to use this endpoint in the laboratory by means of leaf litter decomposition is restricted to species of arable land, e.g. *Lumbricus terrestris* or *Lumbricus rubellus*. Ma (1983) found a similar sensitivity of cocoon production and leaf litter breakdown in *Lumbricus rubellus* in the laboratory. The possibilities to use this endpoint in routine testing are limited. It might be difficult to control the dose of the chemical taken up by the earthworms. As decomposition is a complex process with participation of several faunal groups, this endpoint should preferably be used in field studies.

Outlook

Mortality seems to be a suitable endpoint for classification of chemicals concerning acute effects. The determination of body weight development in the established acute test according to OECD-guideline No. 207 is considered as not sufficient for predicting sublethal effects and longterm exposure of earthworms in arable land. Reproduction and body weight are closely connected endpoints. According to the present state of knowledge both of them should be investigated. The ecological relevance of sublethal effects found in the laboratory needs further investigation.

References

BIERI, M., SCHWEIZER, H., CHRISTENSEN, K. DANIEL, O. (1989) The effect of metaldehyd and methiocarb slug pellets on *Lumbricus terrestris* LINNÉ. *BCPC Monographs, 41; Slugs and snails in world agriculture,* 237–244.

DREWES, C.D., VINING, C.D. (1984) In vivo neurotoxic effects of dieldrin on giant nerve fibers and escape reflex function in the earthworm, *Eisenia foetida. Pesticide Biochemistry and Physiology,* **22,** 93–103.

DREWES, C.D., ZORAN, M.J., CALLAHAN, C.A. (1987) Sublethal neurotoxic effects of the fungicide benomyl on earthworms (*Eisenia fetida*). *Pesticide Science,* **19,** 197–208.

EEC (1985) EEC Directive 79/831. Annex V.Part C: Methods for the determination of ecotoxicity. Level 1. Toxicity for earthworms. Artificial soil test. DG XI/128/82.

EDWARDS, C.A. (1983) Development of a standardized laboratory method for assessing the toxicity of chemical substances to earthworms. *Commision of the European Communities*. Doc EUR 8714 EN. 141 pp.

EDWARDS, C.A. (1984) Report of the second stage in development of a standardized laboratory method for assessing the toxicity of chemical substances to earthworms. *Commission of the European Communities* Doc. EUR 9360 EN, 99 pp.

HAQUE, A., EBING, W. (1983) Toxicity determination of pesticides to earthworms in the soil substrate. *Zeitschrift für Pflanzenkrankheiten und Pflanzenschutz,* **90,** 395–408.

HARTENSTEIN, R., NEUHAUSER, E.F., KAPLAN, D.L. (1979) Reproductive potential of the earthworm *Eisenia foetida. Oecologia,* **43,** 329–340.

HEIMBACH, F. (1985) Comparison of laboratory methods, using *Eisenia fetida* and *Lumbricus terrestris* for the assessment of the hazard of chemicals to earthworms. *Zeitschrift für Pflanzenkrankheiten und Pflanzenschutz,* **92,** 186–193.

LOFS-HOLMIN, A. (1982) Reproduction and growth of common arable land and pasture species of earthworms (Lumbricidae) in laboratory cultures. *Swedish Journal of Agricultural Research,* **13,** 31–37.

MA, W. (1983) Sublethal toxic effects of copper on growth, reproduction and litter breakdown activity in the earthworm *Lumbricus rubellus*, with observations on the influence of temperature and soil pH. *Environmental Pollution* (Series A), **33,** 207–219.

MARTIN, N.A. (1986) Toxicity of pesticides to Allolobophora caliginosa. *New Zealand Journal of Agricultural Research*, **29**, 699–706.

NEUHAUSER, E.F., CALLAHAN, C.A. (1990) Growth and reproduction of the earthworm *Eisenia fetida* exposed to sublethal concentrations of organic chemicals. *Soil Biology and Biochemistry,* **22,** 174–179.

OECD (1984) Guideline for testing of chemicals, No. 207. Earthworm, acute toxicity tests. Adopted 4 April 1984.

PIZL, C. (1988) Interactions between earthworms and herbicides. I. Toxicity of some herbicides to earthworms in laboratory tests. *Pedobiologia,* **32,** 227–232.

STENERSEN, J. (1979) Action of pesticides to earthworms.Part I: The toxicity of Cholinesterase-inhibiting insecticides to earthworms as evaluated by laboratory tests. *Pesticide Science,* **10,** 66–74.

STRINGER, A., WRIGHT, M.A. (1976) The toxicity of benomyl and some related 2–substituted benzimidazoles to the earthworm *Lumbricus terrestris. Pesticide Science,* **7,** 459–464.

VAN GESTEL, C.A.M, VAN DIS, W.A., VAN BREEMEN, E.M., SPARENBURG, P.M. (1988) Comparison of two methods for determining the viability of cocoons produced in earthworm toxicity experiments. *Pedobiologia,* **32,** 367–371.

VAN GESTEL, C.A.M., VAN DIS, W.A., VAN BREEMEN, E.M., SPARENBURG, P.M. (1989) Development of a standardized reproduction toxicity test with the earthworm species *Eisenia fetida andrei* using copper, pentachlorphenol and 2,4,-dichloraniline. *Ecotoxicology and Environmental Safety,* **18,** 305–312.

Use of Earthworms in Eco-neurotoxicity Testing: Sublethal Effects of Carbofuran in *Lumbricus terrestris*

C.D. DREWES AND A. LINGAMNENI

Zoology and Genetics, Iowa State Univ., Ames, IA 50011, USA
Biomedical Engineering, Iowa State Univ., Ames, IA 50011, USA

Abstract

Sublethal neurophysiological effects of the agricultural pesticide, carbofuran, on escape reflex function were studied in intact *Lumbricus terrestris*. Worms were exposed cutaneously to pure carbofuran at dosages comparable to a single commercial granule of Furadan 15G. Neural effects (assessed after 1 h using non-invasive, electrophysiological recordings) included: reduced impulse conduction velocity in medial and lateral giant nerve fibres, and increased delay or failure of coupling between medial giant fibre and motor neuron spikes. Although spasticity and ataxia occurred after 15–20 min, no localized, cutaneous irritation was evident. Ecological relevance of these effects, modes of carbofuran action, and utility of this approach are discussed.

Introduction

Use of the earthworm, *Eisenia foetida*, in conjunction with standardized protocols for chemical exposure (Goats & Edwards, 1988; Heimbach, 1988), has provided an important ecotoxicological tool for predicting field toxicity and interlaboratory and interagency comparisons of relative toxicity of pollutants. Additional efforts to incorporate sublethal parameters as end-points into these tests (Drewes *et al.*, 1987, 1988; Venter & Reinecke, 1988) have increased our understanding of the diverse range of biological effects of toxicants. We cannot expect, however, that such tests always will provide complete and accurate prediction of effects for other earthworm species within different ecological contexts. Therefore, in trying to understand potential biological hazards of pollutants within a particular ecological setting, it seems appropriate and desirable that additional tests be employed, tests in which selection of species and at least some exposure protocols are more closely matched to environmental contexts. Here, we utilize test protocols that enable quantitation of sublethal neurophysiological effects of the agricultural pesticide, carbofuran, on *Lumbricus terrestris*, a prevalent species in many agricultural and urban settings.

Rationale for the experimental approach

PESTICIDE SELECTION

Carbofuran (Furadan) is a widely used agricultural insecticide (Eisler, 1985). For crops such as field corn (i.e., corn rootworm and European corn borer control), it is often applied at planting as covered bands or in seed furrows using a variety of mechanical applicators and one of several granular formulations. Granules are also broadcast as foliar applications for corn borer control. For other crops, such as soybeans, granular Furadan is used for insect and nematode control by applying at planting and mixing with covering soil.

SPECIES SELECTION

Lumbricus terrestris, and other coexistent earthworm species, are broadly distributed in agricultural and urban soils throughout the United States (Reynolds, 1977). However, *L. terrestris* is behaviorally distinguished from other species by vertical burrowing and nocturnal surfacing, behaviours that facilitate successful foraging and mating. In the present study, sexually mature *L. terrestris* (2·8–5·2 g) were maintained at 15°C in closed boxes containing a mixture of horse manure and soil. Before testing, worms were maintained for 24 h at room temperature in covered boxes containing only moist paper towels.

TREATMENT PROTOCOL

In view of agricultural application methods for Furadan and the surfacing behaviour of *L. terrestris*, we used short-term and point-source exposures to either a single commercial granule (Furadan 15G), or an equivalent, pre-weighed amount of pure carbofuran applied to the dorsal skin surface of individual earthworms. The size of a commercial granule was highly variable, but typically within a weight range of 350–1000 μg and diameter range of 0·5–1·0 mm. Two different dosages of pure carbofuran were used: 50 μg and 150 μg, values that correspond closely to the 15% carbofuran-content of typical, small and large Furadan 15G granules.

The site of application was approximately 10–20 segments posterior to the clitellum, corresponding to the center of the region from which electro-physiological recordings were readily obtained. In control worms, a sand grain approximately equal in size to the chemical granule was placed in the same location on the worm's skin. The exposure period was 1 h (21°C). To ensure that the granules stayed in continuous contact with the skin, worms were placed on moist filter paper in a covered dish in dim light. This helped minimize crawling and searching behaviours, which tend to cause inadvertent loss of skin contact with the granule. After exposure and before testing, worms were rinsed briefly in distilled water.

Electrophysiological testing

The medial and lateral giant nerve fibres of earthworms mediate startle withdrawal behaviours that are vital for avoiding predators. Non-invasive recordings of all-or-none action potentials (spikes) from these fibres are readily obtained from a wide variety of earthworm species, including *L. terrestris* (Drewes, 1984).

Giant fibre spikes were initiated by light touch to either end of the worm with a fire-polished glass probe. Electrode grids and preamplification were as previously described (Drewes *et al.*, 1987). Evoked spikes were displayed as internally triggered, single-sweeps on a Tektronix 2221 digital oscilloscope. For most analyses, two-channel recordings were obtained from sites 20 mm apart near the geometric middle of the worm. In treated worms, recording sites straddled the original site of chemical application. Permanent records of results were obtained by transmitting the digitized data from the oscilloscope to a Tektronix HC100 plotter. In experiments requiring higher resolution of the spatial effects of carbofuran treatment, eight-channel, analog recordings were obtained using a Tektronix 5113 storage oscilloscope and two four-channel amplifiers (5A14N). Stored displays were then photographed with a polaroid camera.

Three neural parameters were measured: medial giant fibre (MGF) conduction velocity, lateral giant fibre (LGF) conduction velocity, and time delay from the MGF spike to giant motor neuron (GMN) spike. The latter time is essentially a measure of synaptic delay at the central synapse between MGF and GMN (Drewes, 1984). When determining conduction velocity, care was taken to ensure that body stretch in the recording region was similar for both pre-treatment and post-treatment measurements. Approximately 8–10 replicate measurements were obtained from each worm before, and again after, treatment. Pre-treatment and post-treatment means were then determined for each worm and expressed as relative values, with pre-treatment control values for each worm assigned as 1·0. Usually, 10–13 replicate measurements (to the nearest 0·01ms) of the delay time from MGF-to-GMN spikes were obtained from each worm. Statistical comparisons of relative velocities and delay times within each test group were made using a paired difference t-test. Eight worms were used in each of three test groups: 0 μg-control, 50 μg-treated, and 150 μg-treated.

Behavioral effects

Behavioral effects accompanying treatment with a single commercial granule of Furadan were generally indistinguishable from the effects of pure carbofuran. In all treated worms there were no obvious signs of intoxication or cutaneous irritation during the first 15 min of exposure. Beginning 15–20 min after exposure, however, all treated worms exhibited frequently repeated episodes of whole-body spasms and tight coiling, followed by partial relaxation. Such episodes occurred either spontaneously or in response to slight tactile stimulation. No coordinated or directed locomotion occurred during these episodes (i.e., worms were essentially paralyzed). Similar symptoms were seen in earthworms (*L. terrestris* and *E. foetida*) injected with carbofuran or exposed to carbofuran-treated soils (Stenersen *et al.*, 1973). An important result noted by Gilman & Vardanis (1974) was that *E. foetida*, but not *L. terrestris*, burrowed effectively after injection and appeared to behaviorally discriminate between carbofuran-treated and untreated soils.

Effects on giant fibre conduction velocity

MEDIAL GIANT FIBRE RESPONSES

A typical two-channel recording of a single-spike MGF response to a tactile stimulus, before treatment, is shown in Fig. 1. As previously noted (Drewes,

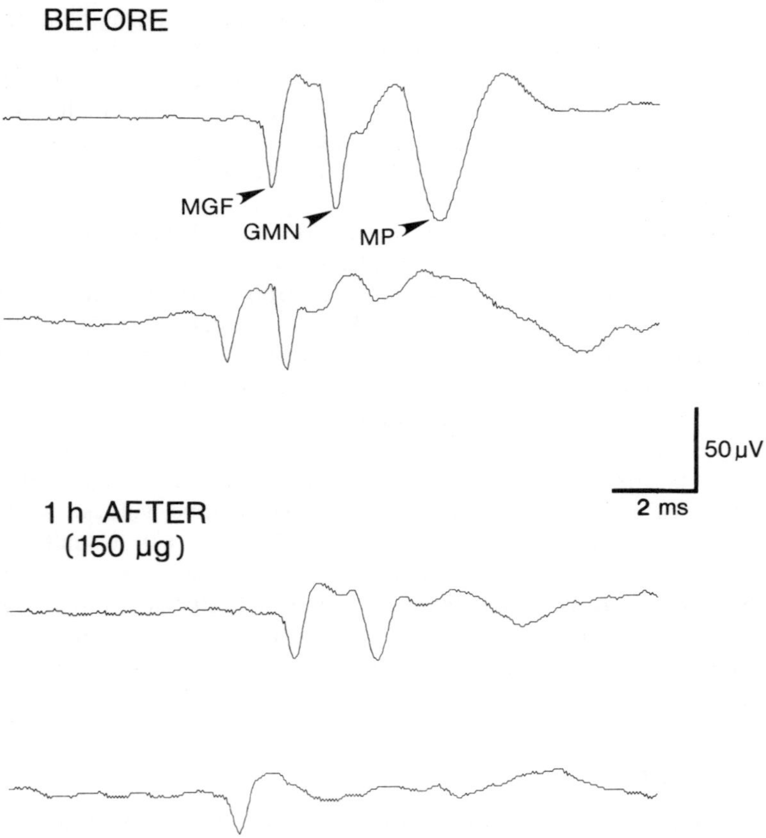

Figure 1. Before treatment, recordings from two sites (upper-posterior, lower-anterior) showed the three components in the MGF response: MGF spike, GMN spike, and muscle potential (MP). One hour after treatment, only the MGF spike was detected at the anterior recording site, while all three components were detected at the posterior site.

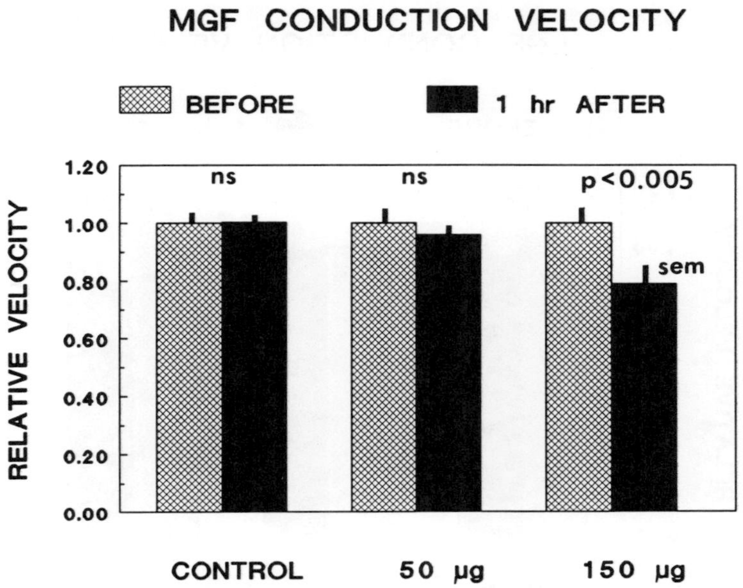

Figure 2. MGF conduction velocity was significantly reduced after 1 h only in the 150 µg-treated group; changes in velocity were not significant (ns) in other groups.

1984), this response has three distinct components: an all-or-none MGF spike, a compound GMN spike, and a small muscle potential. In the control group these waveforms and the relative MGF velocity, were unchanged when worms were retested after 1 h.

No statistically significant change in MGF velocity was noted in the 50 µg-treated group (Fig. 2). However, a significant reduction in velocity, to approximately 80% of pre-treatment, was noted in the 150 µg-treated group. In addition, in four of eight worms in this group, the GMN spike and accompanying muscle potential were occasionally absent in waveforms obtained from one or both recording sites (Fig. 1).

LATERAL GIANT FIBRE RESPONSES

All-or-none LGF spikes were readily detected in both control and treated groups at one hour after exposure. There were no changes in relative LGF velocity in control and 50 µg-treated groups, although a slight reduction was noted in the 150 µg-treated group (Fig. 3).

Effects on MGF-to-giant motor neuron coupling

The delay time from MGF to GMN spikes was unchanged after 1 h in the control group, but markedly increased in carbofuran-treated groups. For example, in the 150 µg-treated group this increase was about 35%, and in the 50 µg-treated group about 15%, after 1 h exposure (Fig. 4).

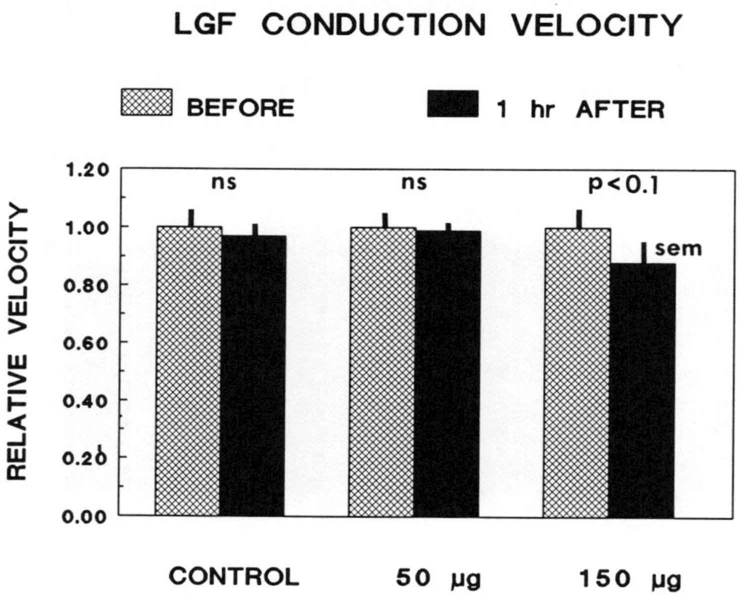

Figure 3. LGF velocity was reduced after 1 h in the 150 µg-treated group.

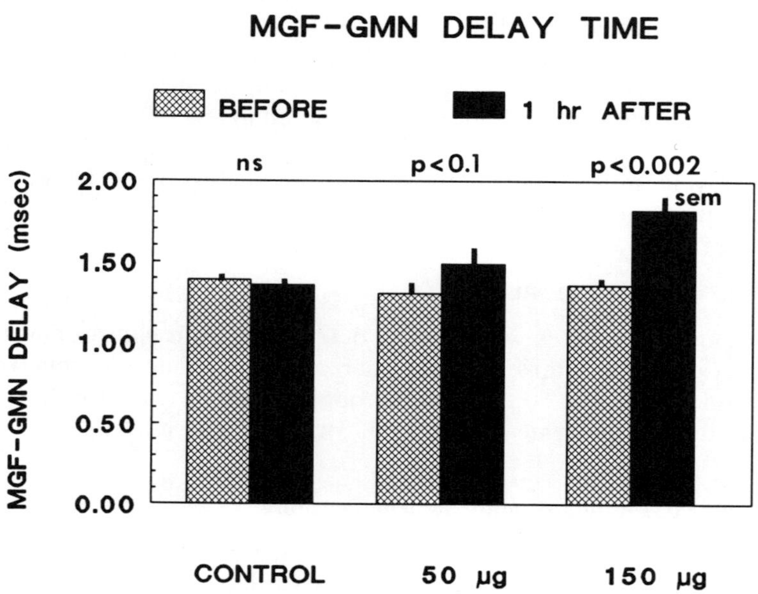

Figure 4. Concentration-dependent increases in the delay time from MGF to GMN spikes were seen in the two treated groups.

To test whether effects were spatially uniform or highly localized within the midbody, simultaneous recordings were obtained from eight different sites spaced 4 mm apart along the midbody of the worm (Fig. 5). Such recordings showed similar reductions in MGF-GMN delay times at each of the sites, suggesting that neural effects of carbofuran are relatively uniform, or systemic, rather than localized to the treatment site.

Possible sites/modes of action

Previous studies by Stenersen (1979 a,b,c) have demonstrated that carbofuran treatment in earthworms is accompanied by cholinesterase inhibition. Since neuromuscular transmission in earthworms is probably cholinergic (Gerschenfeld, 1973), it is possible that peripheral effects at neuromuscular synapses may underly some of the behavioral deficits that accompany carbofuran treatment. Our results, though not refuting this idea, indicate that carbofuran treatment is also accompanied by effects within the worm's central nervous system. These effects include reduced conduction velocity in giant nerve fibres and reduced efficacy of synaptic coupling between the medial giant fibre and its associated giant motor neurons. The latter effect raises the

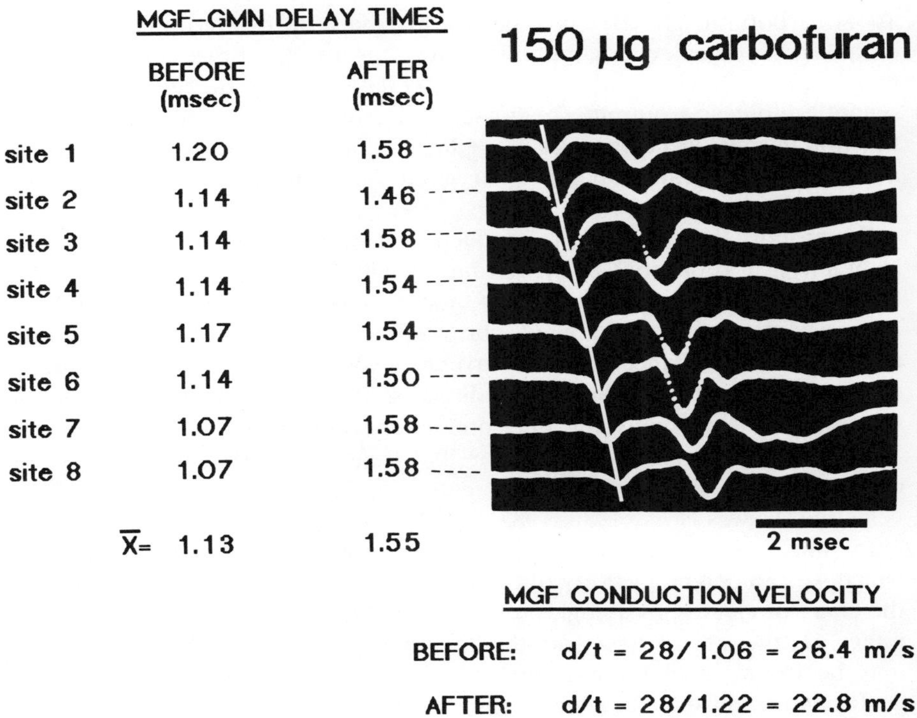

Figure 5. An eight-channel recording showed that the reduction in MGF velocity and the increase in MGF-to-GMN delay were uniformly distributed throughout the midbody region, rather than restricted to one or two recording sites.

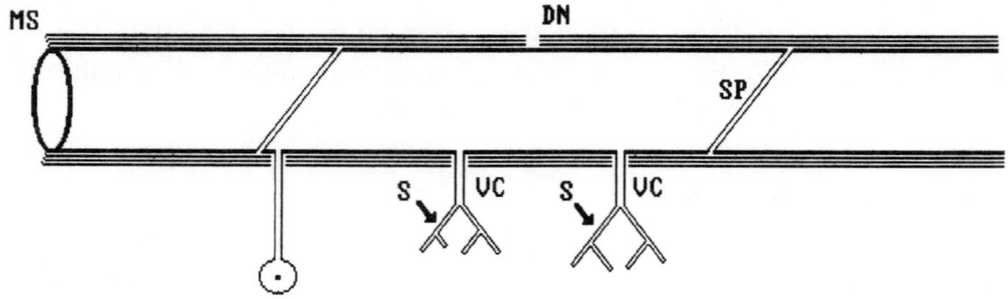

Figure 6. Slowing of giant fibre conduction velocity could occur by several cellular mechanisms, including: 1) decreased membrane resistance of ventral collaterals (VC), due to persisting effects of chemical synaptic inputs (S), 2) altered properties of voltage-sensitive spike channels at the dorsal nodes (DN) that interrupt the myelin-like sheath (MS), or 3) increased resistance of intersegmental, septal partitions (SP).

possibility that transmission at this synapse has a cholinergic component, but the question is open since supporting pharmacological data is non-existent.

The reduction in giant fibre conduction velocity raises additional questions about carbofuran effects. Since earthworm giant fibres are multisegmental, each segment has several potential sites for toxic actions on giant fibre membranes. These include: (a) ventral collaterals that function as sites of synaptic input and output, (b) dorsal nodes in the myelin-like sheath that function in spike electrogenesis in the MGF, and (c) gap junctional partitions at the intersegmental boundary of giant fibres in some segments (c.f., Drewes, 1984).

We propose that decreases in conduction velocity could result from changes in biophysical properties of membranes at any of these sites. For example, treatment-induced accumulation of neurotransmitter (e.g., acetylcholine) near synaptic inputs onto giant fibre collaterals might increase collateral membrane conductance so that currents associated with giant fibre spike conduction would be 'short-circuited' across these sites, thereby decreasing the effective space constant and conduction velocity of the fibre (Fig. 6). Whatever the mechanism, it is clear that treatment with a wide variety of compounds is associated with decreases in giant nerve fibre conduction velocity (Drewes *et al.*, 1986, 1987).

In summary, our approach provides an 'experimental window' through which we can detect a variety of treatment-induced lesions or 'weak links' in the chain of electrophysiological events that underly rapid escape reflexes in intact earthworms. Effects on the functional properties of these reflexes may be useful in two ways. First, they may be indicative of the general performance, during and after treatment, of synaptic transmission within the intact central nervous system. Second, sublethal neurotoxic effects on escape reflex function may have implications in terms of increased susceptibility to predation.

Ecological significance

Since cutaneous exposure to granular doses of carbofuran can quickly, and without signs of behavioral avoidance, paralyze an earthworm, it seems important to consider granular application rates in assessing neurotoxic risk. Given that a typical granule of Furadan 15G weighs about 0·7 mg and that recommended application rates may be up to 13·3 lbs/acre, we estimate a concentration of 200 granules/ft^2 near the soil surface. Thus a worm's normal locomotor activities at or near the soil surface would very likely bring it into cutaneous contact with one or more granules. If the effective chemical concentration in each granule was not immediately and substantially reduced by water solubility and diffusion, we assume that neurotoxicity effects would occur, leading to mortality at the soil surface by predation, desiccation, or ultraviolet irradiation. If contact occurred below the soil surface (e.g., by exposure to soil leachates), the question of persistence or recovery from neural effects becomes critical for restoring earthworms' contributions to soil fertility.

Acknowledgements

We thank Dr. E. Berry and A. Swalla for providing facilities, assistance and earthworms. Some of these experiments were carried out in the National Soil Tilth Laboratory, Ames, IA.

References

DREWES, C.D. (1984) Escape reflexes in earthworms and other annelids. In: *Neural Mechanisms of Startle Behaviour*, R.C. Eaton (Ed.), New York: Plenum, pp. 43–91.

DREWES, C.D., ZORAN, M.J., CALLAHAN, C.A. (1987) Sublethal neurotoxic effects of the fungicide Benomyl on earthworms (*Eisenia foetida*). *Pesticide Science,* **19,** 197–208.

DREWES, C.D., VINING, E.P., CALLAHAN, C.A. (1988) Electrophysiological detection of sublethal neurotoxic effects in intact earthworms. In: *Earthworms in Waste and Environmental Management*, C.A. Edwards and E.F. Neuhauser (Eds.), The Hague: SPB Academic Publishing, pp. 355–366.

EISLER, R. (1985) Carbofuran hazards to fish, wildlife, and invertebrates: a synoptic review. United States Department of Interior/Fish and Wildlife Service, *Biological Report 85(1·3), Contaminant Hazard Reviews Report No. 3*, pp. 1–36.

GILMAN, A.P., VARDANIS, A. (1974) Carbofuran. Comparative toxicity and metabolism in the worms *Lumbricus terrestris L.* and *Eisenia foetida S. Journal of Agricultural and Food Chemistry,* **22,** 625–628.

GOATS, G.C., EDWARDS, C.A. (1988) The prediction of field toxicity of chemicals to earthworms by laboratory methods. In: *Earthworms in Waste and Environmental Management*, C.A. Edwards and E.F. Neuhauser (Eds.), The Hague: SPB Academic Publishing, pp. 283–294.

HEIMBACH, F. (1988) A comparison of laboratory methods for the toxicity testing of earthworms. In: *Earthworms in Waste and Environmental Management*, C.A. Edwards and E.F. Neuhauser (Eds.), The Hague: SPB Academic Publishing, pp. 329–335.

REYNOLDS, J.W. (1977) The Earthworms (Lumbricidae and Sparganophilidae) of Ontario. Toronto: The Royal Ontario Museum, pp. 99–103.

STENERSEN, J. (1979a) Action of pesticides on earthworms. Part I: The toxicity of cholinesterase-inhibiting insecticides to earthworms as evaluated by laboratory tests. *Pesticide Science,* **10,** 66–74.

STENERSEN, J. (1979b) Action of pesticides on earthworms. Part II: Elimination of parathion by the earthworm *Eisenia foetida* (Savigny). *Pesticide Science,* **10,** 104–112.

STENERSEN, J. (1979c) Action of pesticides on earthworms. Part III: Inhibition and reactivation of cholinesterases in *Eisenia foetida* (Savigny) after treatment with cholinesterase-inhibiting insecticides. *Pesticide Science,* **10,** 113–122.

STENERSEN, J., GILMAN, A., VARDANIS, A. (1973) Carbofuran: Its toxicity to and metabolism by earthworm (*Lumbricus terrestris*). *Journal of Agricultural and Food Chemistry,* **21,** 166–171.

VENTER, J.M., REINECKE, A.J. (1988) Sublethal ecotoxicological effects of dieldrin on the earthworm *Eisenia foetida* (Oligochaeta). In: Earthworms in Waste and Environmental Management, C.A. Edwards and E.F. Neuhauser (Eds.), The Hague: SPB Academic Publishing, pp. 337–353.

Measuring the effects of chemicals on earthworms in the field

Testing the Effects of Chemicals on Earthworms: The Advantages and Limitations of Field Tests

C.A. EDWARDS

The Ohio State Unversity, Columbus, Ohio, USA

Introduction

The soil is one of our most valuable natural resources, so the maintenance of soil fertility is of major importance. The main factor promoting soil fertility is the breakdown of dead organic matter and the release of the nutrients that it contains. The main agents in these processes are the multitudes of soil organisms that are characteristic of healthy soils. Hence, it is extremely important that populations of soil-inhabiting micro-organisms and invertebrates be maintained. Earthworms are probably the most important soil-inhabiting invertebrates in promoting soil fertility (Edwards and Heath, 1964).

Some pesticides and industrial chemicals can have drastic effects on soil organisms; hence it is important to know the relative toxicity of chemicals that may reach soils, so that the exposure of soil organisms to the more toxic chemicals can be avoided or minimised. The various organisms that inhabit soils differ greatly in their susceptibility to chemicals (Edwards, 1973; Edwards and Thompson, 1973; Thompson and Edwards, 1974) so it is very difficult to assess the overall effects of different chemicals on dynamic soil processes that are mediated by soil organisms. Earthworms are extremely important in the breakdown of organic matter into soil, the release of nutrients, the aeration and drainage of soil and the formation of soil aggregates.

Earthworms have a number of characteristics which identify them as one of the most suitable soil animals to be used as a key bioindicator organism to assess soil pollution by chemicals. In addition to their key role in soil fertility, they are common in the great majority of soils, they are large in size and easy to handle, they can be readily collected and identified and are known to be affected by and take up into their tissues, a number of organic and inorganic chemicals.

The effects of chemicals on earthworms and methods of investigating these effects have been reviewed by Satchell (1955), Davey (1963), Edwards and Thompson (1973), Thompson and Edwards 1974) and Lofs-Holmin and Boström (1988), Edwards and Neuhauser (1988).

Most of the earlier reports in the scientific literature refer to field testing of the effects of chemicals, particularly pesticides, on earthworm populations. These are too numerous to list (Edwards and Thompson, 1973) but range from the effects of lead arsenate (Polivka, 1951; Escritt, 1955) copper sulphate and organochlorines (Davey, 1963; Edwards *et al*, 1967) carbamate compounds (Thompson, 1971; organophosphorus compounds (Edwards *et al*, 1968; Thompson, 1970; Way and Scopes, 1968) and herbicides (Edwards, 1970). When the results of these field experiments are reviewed it is extremely difficult to assess reliably the relative toxicity of different pesticides, due to considerable variability between test sites, soils, formulations, doses and methods of application used. The field tests reported in the literature are adequate to identify chemicals which are extremely toxic to earthworms but do not identify accurately, moderately or slightly toxic compounds (Lofs-Holmin and Boström, 1988; Edwards and Bohlen, in press). There are at least 250 reports in the literature on the testing of effects of chemicals on earthworms, either alone or in combination with other organisms in field tests. These tests have been extremely variable in design and execution and the results are extremely difficult to compare and assess (Lofs-Holmin and Boström, 1988; Edwards and Bohlen, in press) in terms of potential environmental hazard.

There are many methods of testing the effects of chemicals on earthworms in the laboratory and one of these, the artificial soil test (Edwards, 1983, 1984; Goats and Edwards, 1988) has been adopted by the EEC and OECD as a standard test. However, there is an urgent need for a standardised field test so that laboratory and field test results can be correlated, with the ultimate aim of using the laboratory test for hazard prediction without the need for expensive field tests.

The aim of this paper is to review the advantages and limitations of field tests for assessing the toxicity of chemicals to earthworms and to make suggestions on a standardised design for such field experiments.

Advantages of Field Tests for Assessing Toxicity of Chemicals to Earthworms

Most of the earliest reports, in the literature of the effects of chemicals on earthworms, involved pesticides and were aimed primarily at assessing the effects of these chemicals on earthworm populations, with the overall objective of avoiding the use of pesticides that were very toxic to earthworms.

Field tests that assess the toxicity of chemicals to earthworms have some obvious advantages. The exposure of earthworms to the chemicals is identical to that to which they are exposed after the application of agricultural chemicals to soils or when industrial chemicals fall on the surface of the soils. Moreover, the exposure of the earthworm to chemicals is under natural climatic and environmental conditions. In field tests a range of development stages of earthworms are exposed simultaneously to chemicals and the toxic effects of the chemical are superimposed on natural mortality and predation. By comparison, most laboratory tests involve a single life stage of earthworm, natural mortality is minimised and the test worms are in optimal health. In the field a range of species of earthworms are exposed to chemicals simultaneously, compared to the single species used in laboratory tests that may not be totally representative of the general population. In field tests, if a sequential series of samples is taken in time, the longevity of the toxic effects of the chemical under field conditions can be assessed, compared with a relatively short period of exposure to chemicals that is possible under the standardised conditions in the laboratory.

Most laboratory tests are designed to assess only the acute or chronic direct toxicity of a chemical. By comparison, field tests can also evaluate the indirect effects of chemicals. These include: effects of the chemical on the earthworm's food supply, eg. herbicides affect the amount of decaying plant material in soil, and fungicides may influence populations of fungi that are a major component of the diet of many species of earthworm (Edwards and Fletcher, 1988). Similarly, chemicals can have indirect effects on the earthworm habitat, eg. herbicides affect the degree of soil cover which keeps the surface soil much moister.

Earthworms can take up many chemicals from soil into their tissues, and in field experiments the exposure is under a range of different environmental conditions. By comparison, the exposure in the laboratory can only be under selected constant conditions. For example, in the field, aldicarb is taken up into earthworm tissue only in the form of the parent compound in waterlogged soils and not as the degradation product. Such a type of differential uptake would not be determined in laboratory experiments under controlled environmental conditions.

Clearly, field experiments have many advantages in the assessment of potential hazards from chemicals but there is an urgent need for a standardised design of field experiments and defined criteria for assessing results if they are to be effective.

Limitations of Field Tests for Assessing Toxicity of Chemicals to Earthworms

FACTORS THAT INFLUENCE THE EXPOSURE OF EARTHWORMS TO CHEMICALS

The amount of exposure of earthworms to chemicals in soils depends upon a wide range of variable *factors associated with the chemical*, the soil type, the environmental conditions, and the behaviour of the earthworms. Chemical characteristics that affect their impact upon earthworms. Chemical characteristics that affect their impact upon earthworms include: volatility, water solubility, lipid/water partition coefficient, adsorptive capability and persistence in soils.

Various *characteristics of soils* can influence the toxicity to earthworms of chemicals that reach it. When soil has a low moisture content, the toxicity of some chemicals can be decreased through adsorption on to soil fractions, whereas compounds that dissolve readily in water can be leached from a soil that is water-logged. Chemicals can bind reversibly or irreversibly with the particles of clay and organic matter in the soil, and the proportions of these two components in soil will also influence the cation exchange capacity of the soil, which will affect the amount of chemical that is available to the earthworms. Soil temperature affects the vapour pressure of a chemical and also its solubility in soil water, thereby influencing the rate at which compounds are taken up by earthworms. The rates of loss of chemicals from the soil as a vapour or through the action of soil micro-organisms also depends upon soil temperature. Micro-organisms are instrumental in detoxification and degradation of many chemicals, yet the range of detoxification and degradation of many chemicals, yet the range of micro-organisms present in the soil at an experimental site, and their subsequent activity, is seldom known because it is difficult to determine and standardise.

The *behaviour of earthworms* can affect their exposure to chemicals in the field. The horizontal distribution of earthworms in the soil is limited by certain aspects of the physicochemical environment, such as the temperature, moisture content and pH of the soil, the availability of food and the capacity of the earthworms to reproduce and disperse. Vertical distributions of earthworms are also affected by changes in the soil conditions, such as temperature and moisture (Edwards and Lofty, 1977) and the ability of the earthworms to respond to these changes. Juvenile earthworms are often unable to burrow deeply in the soil and are therefore affected more severely than the adults by toxic chemicals. By contrast, adult earthworms of deep burrowing species can move rapidly into the safety of the deep soil layers using permanent, mucous-lined burrows. Earthworms can move into diapause under adverse temperatures or moisture conditions.

Different earthworm species can be exposed to quite different amounts of chemicals in different ways. For instance, *L. terrestris* is often exposed to a high concentration of pesticides, because this species of earthworm moves on and feeds on the soil surface. *A. caliginosa* lives in the superficial layers of the soil and the adults move on the soil surface in wet weather, becoming particularly vulnerable to surface pesticide residues. *A. longa* seems to be less susceptible to pesticides than many other species of earthworm, because it can burrown deep into the soil (Wheatley and Hardman, 1968) and enters an obligatory diapause during the summer (Gerard, 1967; Edwards and Lofty, 1977).

Both *L. terrestris* and *A. longa* construct permanent burrows (Edwards, 1983) and this probably decreases the exposure of these earthworms to pesticides that are incorporated into the soil. However, under some conditions such burrows can channel chemicals that are drenched onto the soil, allowing them to percolate rapidly and make rapid contact with the earthworms.

Earthworm activity affects exposure. For instance, temperate earthworm species have a peak period of activity from April to May and from August until early December. *L. terrestris* appears to become inactive in summer (Gerard, 1967), and *A. longa* enters an obligatory diapause from May until

October, so *L. terrestris* and *A. longa* have similar seasonal behaviour patterns. *A. caliginosa, A. chlorotica* and *A. rosea* can enter a facultative diapause during unfavourable conditions (Evans, 1947) and such resting behaviour can affect results of field tests. The activity of earthworms in soil is limited mainly by temperature and moisture, although drought induces a more profound quiescence than winter cold (Gerard, 1967). The optimal temperature for earthworm activity can differ between species. *A. caliginosa* and *E. fetida* are most active in soils between 10–23°C and 16–23°C respectively, whilst *L. terrestris* is more active in soil at 10°C (Edwards and Lofty, 1977). High temperatures tend to limit the activity of earthworms more than low temperatures. The removal of an insulating layer of vegetation from the field cultivations can increase the range within which the temperature of the soil fluctuates, and thus affect populations of earthworms adversely (Edwards, 1984).

VARIABILITY IN FIELD EXPERIMENTS ON THE TOXICITY OF CHEMICALS TO EARTHWORMS

The field experiments which have assessed the effects of chemicals on earthworms in the literature are extremely diverse and poorly-designed (Lofs-Holmin and Boström, 1988). As many as 7% of field experiments had *no control or check* plots other than by a loose comparison with earthworm populations in fields close to the one treated with chemicals. Many experiments in the literature have inadequate replications which would enable statistical analyses of the data to be made.

More than a third of the field experiments reported in the literature either had *no replicates* (12%) or gave inadequate information on the degree of replication (24%). About half had 2–4 replicates and the rest had 5-10 replicates. Nearly a quarter of the field experiments did not define *the dose*, other than to state that it was ‘recommended’, ‘commonly-used’ or ‘normal’.

Little information on the *soil type* or its characteristics was given in most field experiments. Less than half of them reported the soil pH, texture or organic matter content. Very little information on *climatic variation* during the exposure period was given and the periods of exposure to the chemical were very variable, most lasting 1–6 months but some continued for up to 5 years. A few of the experiments reported lasted less than one month.

The effects of chemicals on earthworms was assessed using *extremely variable critera*. These include: assessment of earthworm populations (as numbers or biomas) by hand-sorting, pitfall traps, formalin extraction, counting dead earthworms on the soil surface, counting earthworm casts, measuring the concentration of the chemical in earthworms, assessing the consumption of organic matter from the surface, or counting earthworm burrows. Some of these indices are affected by seasonal activity, the behaviour of earthworms and a variety of environmental factors.

Clearly, if field tests for assessing the toxicity of chemicals to earthworms are to be useful and comparable, there must be much more standardisation in their design and procedures.

The Design of Standardised Field Experiments

There is bound to be some variability in the results of field experiments, due to differences in soil type, climatic factors and species and numbers of earthworms present. However, field experiments can provide valuable data that is almost impossible to obtain from laboratory experiments, if a carefully standardised design of field experiment is used.

The following suggestions for the design of field experiments to assess the toxicity of chemicals to earthworms in the field can be made.

SITE VARIABLES

It is important to choose a site carefully. The important factors influencing this choice are:

– A large population of earthworms with a representative range of *at least six species of common earthworms*. Preferably at least 100 earthworms per m^2 should be present prior to application of the chemical, although a lower population does not preclude the sue of a site; it means that more samples will be needed. Such populations can be found in many arable soils that have not been exposed to pesticide treatments, in temperate regions. *Lumbricas terrestris* is probably the most important species, because it incorporates organic matter into soil, it has permanent burrows through which chemicals can percolate, and it becomes exposed to surface residues of chemicals more often than other species, so the test site should have this species if possible. *A. caliginosa* is probably the most important shallow-working species and should preferably be present.

– The *soil-type* should be preferably loam-based, avoiding extreme clay soils which bind up chemicals and sandy soils which do not favour earthworms. The soil type should be as homogeneous as possible across the experimental site. The classification and description of the soil type and its organic matter content should be included in all reports.

– As far as possible there should be *no history of chemical use on the site for at least 5 years* prior to the experiment. An analysis of the soil for chemical residues prior to the experiment is recommended.

TREATMENT VARIABLES

The way in which the chemical reaches the earthworm influences its effect greatly:

– The *chemical* should be clearly defined in terms of structure and mix of isomers. If it is a pesticide its common name should be known. Its physicochemical properties such as solubility, water/lipid partition coefficient and volatility, which can be used in prediction of toxicity should be known. Its formulation which also influences its toxicity should be stated.

– The *dose* used should be relevant to the exposure likely to occur under natural conditions. For agricultural chemicals, it should be the highest recommended dose (kg/ha active ingredients). For industrial chemicals it should parallel amounts reported in environmental surveys. If higher doses are tested they should be in the series 2x, 4x, 8, 16x, the recommended dose, to enable dose/effect curves to be plotted.

– The *chemical should always be applied to replicated plots*. As a general guide a minimum of four replicates should be used. The greater the number of replicates the greater the precision of the experiment.

– The *size of plot* is important. Earthworms, other than *L. terrestris* do not migrate far laterally, so relatively small plots with no physical barriers between them are adequate. Plots should be at least 5 metres square and up to 10 metres square. With plots larger than this, the potential variability of the soil type over a large area can increase the variability of the results. There is considerable evidence that earthworms do not migrate significantly between plots of these sizes (Edwards, 1973 a&b).

– *the cropping of the experimental site should be defined*. If a chemical is tested that would be expected to reach the ground in fall-out or precipitation it would be appropriate to apply the test chemical to grassland that contains a suitable earthworm population. However, if a chemical is to be tested that would normally be applied to arable land, it should be applied to bare soil to avoid binding the pesticide to the surface thatch commonly found in grassland. Preferably, it should be an arable site with a suitable earthworm population. After the chemical has been applied, the plots can then be seeded to grass or other crops. Row crops should be avoided since they complicate the assessment of earthworm populations. The testing of herbicides in the field is more difficult because the indirect effects of the herbicide on earthworms through changed weed populations may be greater than the direct effects of the chemical so, to assess direct toxicity, plots must be kept fallow (Edwards and Stafford, 1978). When an arable site with an adequate earthworm population is not available, an alternative is to cultivate a grassland site and preapre a seedbed before applying the test chemical.

– The *timing of the experiment* is very important. In temperate countries earthworms are most active in spring and autumn. It is recommended that whenever possible, experiments should begin in spring and should all be continued for a minimum of six months to include the autumn months, but longer (up to two years or more) for persistent chemicals.

– It is recommended that whenever possible the effects of a test chemical should be compared with those of a *toxic standard*. Possible standard chemicals include benomyl or carbaryl (Edwards 1983, 1984).

– If the issue of entry into food chains is to be addressed, residues of the chemical in earthworms and soil should be measured to assess whether *bioconcentration* occurs (Edwards, 1973 a&b). Methods of assessing residues in worms are described elsewhere (Edwards, 1979).

ASSESSMENT CRITERIA

The most important criteria in assessing the overall effects of a chemical on earthworms is the *overall effect on populations* (numbers and biomass) of earthworms. The assessment of populations of soil-inhabiting invertebrates such as earthworms is extremely difficult (Edwards, 1991).

The most thorough *method of assessing populations* of earthworms, to include all stages of the population (cocoons, immatures, resting and active adults), is by hand-sorting soil sample from a quadrat (0·25–1·0 m^2) to a depth of 30–60 cm in the field or laboratory. However, this is extremely laborious and time-consuming.

An alternative method of assessing populations in the field is by using quadrates (0·25–1·0 m^2) and pouring litres of dilute formalin (50 cc of 40% formaldehyde in 9 litres of water) on to the soil gradually. Earthworms will come to the surface within 15–20 minutes, can be picked up with forceps and stored in 5% formalin until they can be identified. This method is relatively quick but has the drawback that it works best when earthworms are at peak seasonal activity and is most effective for those earthworms with permanent burrows such as *L. terrestris* and *Allolobophora longa*. This method does not sample cocoons or diapausing worms.

The method recommended for maximum efficiency is to hand-sort earthworms in the top 5–10 cm of soil in the field, clear this soil from the quadrat and then pour dilute formalin on to it. This method recovers both these earthworm species that live most of their lives close to the soil surface (endogeic), as well as those that penetrate as much as 2 metres deep into the soil through burrows (anecic) and therefore would not be recovered by hand-sorting.

It has been suggested that an electrical method of sampling by passing a current from a 220 volt generator through soil is a suitable sampling method (Satchell, 1955; Edwards and Lofty, 1977; Thielemann, 1986). However, this method does not sample a sharply delimitated area of soil and is greatly influenced by soil pH and moisture content.

– The *timing of earthworm sampling* in field experiments should be related to the persistence of test chemicals in soil, as far as it is known. For chemicals that persist less than six months in soil, monthly sampling is adequate. For chemicals that persist longer it is recommended that samples should be taken 1,2,4,8,12 and 24 months after treatment with sampling every six months after that for very persistent chemicals.

– The *number of earthworm population samples* needed per replicate can differ with variations in soil type, poulations of earthworms and size of plot. However, it is recommended that a minimum of four samples per plot be taken on each sampling date if a statistical change in population of the order of 25% is to be identified.

Conclusions

The value of field experiments in assessing the effects of chemicals upon earthworms can be increased enormously by using carefully designed experiments. We need much more data from well-designed and standardised field experiments that can be compared with data for the same chemical from standardised laboratory experiments in order to test the validity of the laboratory test. However, the results should be used whenever possible combined with full knowledge of the physico-chemical characteristics of the compound under test and the results of at least a simple laboratory test (Figure 1) (Edwards, 1990). By combining the results from the field and laboratory it is much easier to reach a balanced judgement on the potential toxicity and environmental hazard of a compound. As a suitable data bank is accumulated involving not only physico-chemical data, but also the results of laboratory and field experiments for chemicals with particular structures and properties, accurate prediction of particular hazards to earthworms should become much more feasible.

References

DAVEY, S.P. (1963). Effect of Chemicals on Earthworms: A Review of the Literature. USDI Special Scientific Report Wildlife No 74, Washington USA, pp 20.

EDWARDS, C.A. (1973a). Environmental Pollution by Pesticides. Plenum Press, London and New York, pp 542.

EDWARDS, C.A. (1973b). Persistent Pesticides in the Environment. 2nd Edition. CRC Press, Cleveland, OH. pp 170.

EDWARDS, C.A. (1979). Tests to Assess the Effects of Pesticides on Beneficial Soil Organisms. Proc. Umweltbundsamt Seminar on Ecological Tests Relevant to Environmental Chemicals: Evaluation and Research Needs. December 1977, 240–253.

EDWARDS, C.A. (1983). Development of a standardised laboratory method for assessing the toxicity of chemical substances to earthworms. Report of Commission of European Communities. EUR.

EDWARDS, C.A. (1984). Report of the second stage in development of a standardised laboratory method for assessing the toxicity of chemical substances to earthworms. Report for Commission of European Communities. EUR 9360 EN. pp 99; 8714 EN. pp 141.

EDWARDS, C.A. (1991). Methods for assessing populations of soil-inhabiting invertebrates. Agriculture, Ecosystems and Environment **34**, 145–176.

EDWARDS,C.A. AND BATER, J.E. (1990). An evaluation of laboratory and field studies for the assessment of the environmental effects and pesticides. British Crop Protection Conference, Pests and Diseases, 963–968.

EDWARDS, C.A. AND BOHLEN, P. (1991). The Assessment of the Effects of Toxic Chemicals Upon Earthworms. In: Reviews of Environmental Contamination and Toxicology. Ed G. Ware. (in press).

EDWARDS, C.A., DENNIS, E.B. AND EMPSON, D.W. (1967). Pesticides and the soil fauna effects of aldrin and DDT in an arable field. Annuals of Applied Biology, **60**, 11–22.

EDWARDS, C.A. AND FLETCHER, K.E. (1988). Interactions Between Earthworms and Microorganisms. Agriculture, Ecosystems and Environment, 235–249.

EDWARDS, C.A. AND LOFTY, J.R. (1977). Biology of Earthworms. 2nd Edition, Chapman and Hall, London, pp 333.

EDWARDS, C.A. AND NEUHAUSER, E.F. (ED.) (1988). Earthworms in Waste and Environmental Management. SPB Academic Publishing, The Hague, The Netherlands, pp 391.

EDWARDS, C.A. AND STAFFORD, C.J. (1978). Interactions Between Herbicides and the Soil Fauna. Annuals of Applied Biology, **89**, (2), 91, 125–146.

EDWARDS, C.A. AND THOMPSON, A.R. (1973). Pesticides and the Soil Fauna. Residue Reviews, **45** (1), 1–73.

EDWARDS, C.A., THOMPSON, A.R. AND LOFTY, J.R. (1968). Changes in soil invertebrate populations caused by some organophosphorus insecticides. Proceeding of 4th British Insecticide and Fungicide Conference, 48–55.

ESCRITT, J.R. (1955). Calcium arsenate for earthworm control. Journal of the Sports Turf Research Institute **9**.

GERARD, B.M. (1960). The biology of certain British earthworms in relation to environmental conditions. PhD Thesis, University of London.

GOATS, G.C. AND EDWARDS, C.A. (1988). The Prediction of Field Toxicity of Chemicals to Earthworms by Laboratory Methods. In: Earthworms in Waste and Environmental Management, Ed. C.A. Edwards and E.F. Neuhauser. SPB Academic Publishing, The Hague, The Netherlands, 283–294.

GOATS, G.C. (1985). Assessment of the Effects of Toxic Chemicals Upon Earthworms. PhD Thesis, University of London, pp 351.

LOFS-HOLMIN, A. AND BOSTRÖM, U. (1988). The Use of Earthworms and Other Soil Animals in Pesticide Testing. In: Earthworms in Waste and Environmental Management. Ed. C.A. Edwards and E.F. Neuhauser, SPB Academic Publishing, The Hague, The Netherlands, 303–313.

POLIVKA, J.B. (1951). Effect of insecticides on earthworm populations. Ohio Journal of Science. **51**, 195–196.

ROBERTS, B.L. AND DOROUGH, W.H. (1984). Relative Toxicities of Chemicals to the Earthworm *Eisenia foetida*. Environmental Toxicology and Chemistry **3**, 66–78.

SATCHELL, J.E. (1955a). The effects of BHC, DDT and parathion on the soil fauna. Soils and Fertilisers. **18**, 279–285.

SATCHELL, J.E. (1955b). An electrical method of sampling earthworm populations. In: Soil Zoology. Ed. D.K. McEkewan, Butterworths, London 356–364.

SAUNDERS, D.G. AND FORGIE, C.D. (1977). Some effects of phorate on earthworm populations. Proceedings 30th New Zealand Weed and Pest Control Conference, 222–226.

THIELEMANN, U. (1986). Elektrischer Regenwurmfans mitt der Oktette-Methode Pedobiologia **29**, 295–302.

THOMPSON, A.R. AND EDWARDS, C.A. (1974). Effects of Pesticides on Non-Target Invertebrates in Freshwater and Soil. In: Pesticides in Soil and Water. Soil Science Society of America Special Publication **8**, 1974. 341.

THOMPSON, A.R. (1971). Effects of nine insecticides on the numbers and biomass of earthworms in pasture. Bulletin of Environmental Contamination and Toxicology **5**, 577–586.

WAY, M.J. AND SCOPES, N.E.A. (1968). Studies on the persistence and effects on soil fauna of some soil-applied systemic insecticides. Annals of Applied Biology **62**, 199–214.

WHEATLEY, G.A. AND HARDMAN, J.A. (1968). Organochlorine residues in earthworms from arable soils. Journal of Science of Food and Agriculture, **19**, 219–225.

Measuring Effects of Pesticides on Earthworms in the Field: Effect Criteria and Endpoints

A. LOFS

Swedish University of Agricultural Sciences, Department of Ecology and Environmental Research, S-750 07 Uppsala, Sweden

Abstract

The influence of pesticides on earthworms has to be measured as a change in abundance and mean weight in at least three stages of development for each species recorded. The endpoints have to relate to natural fluctuations and those caused by other human activities in arable land and grassland, and also to the ability of the populations to recover. A statistically significant reduction of more than 20–30% in abundance of juveniles after one year, compared to controls, should be considered unacceptable.

Introduction

There are two main reasons to investigate toxicity of pesticides on earthworms:

1. Earthworms are of ecological importance in arable fields and grasslands. Thus, there is an economic interest in testing side effects and to quantify them. The contribution of earthworms to organic matter turnover and plant nutrient circulation can be estimated. A decline in number of biomass caused by a pesticide might be significant to the soil/plant ecosystem in relation to natural fluctuations and ability to recover.
2. Earthworms are practical and representative monitoring organisms, ie. in a prolonged sense of ecological thinking: What is bad for earthworms might be bad for humans. In this case the test must detect even slight effects of an environmental hazard disregarding the function of earthworms in soil.

Ecological reasons for population studies

Consumption by earthworms has recently been quantified to be 11–14 times their biomass measured as grams carbon per m^2 in different tilled or non-tilled arable systems in Sweden, or nearly one third of the above-ground litter annually produced (Andrén *et al.*, 1990).

Contribution of earthworms to litter disappearance in a grass ley was estimated to be 10–36% of annual production (Boström, 1988). Thus, their ecological part in litter breakdown and mineralisation process is significant. The economic value is harder to estimate and appreciate except where earthworms have been artificially introduced.

Population fluctuations in arable land

A question which needs to be answered is 'what is an ecologically significant population decrease due to pesticide treatment in view of natural variation between years?' In the 'Ecology of Arable Land', a project in Sweden, it was stated that earthworm populations can fluctuate enormously. The biomass could be reduced by 80% and abundance with 70% during a winter with repeated freezing and subsequent waterlogging (Boström & Lofs-Holmin, 1988).

RECOVERY AND CHANGES IN SPECIES COMPOSITION

Rotary cultivation of a lucerne ley destroyed 70% of the earthworms or 97% of the biomass (Boström, 1988). A year later the earthworm biomass had recovered. This was due to a large pool of surviving cocoons in the soil and a good supply of crop residues ploughed under. The species' composition, however, had drastically changed. Under lucerne, nearly half of the biomass consisted of species other than *A. caliginosa*. A year later that part had declined to only 10%. In a grass ley corresponding figures were 30% before and 20% after cultivation. Despite a total biomass reduction of 85% (60% in numbers), the biomass had not only recovered but a 40% increase was noted one year later compared to undisturbed plots.

A population study which does not distinguish between species can overlook large biological effects. A pesticide can make litter unpalatable for surface feeding species (Cook, 1983), while others might be favoured by increased plant production so that total biomass or number will be unchanged or even increased (Boström & Lofs-Holmin, 1986).

Choice of parameters

The parameter most commonly studied in field tests is abundance. The second most investigated parameter is the concentration of a chemical or its degradation products within earthworms. This is relevant if the pesticide is persistent or frequently used and might enter food chains without harming earthworms. Biomass is the third commonest parameter. Mortality is seldom directly measured. (Lofs-Holmin & Boström, 1988).

Population changes are complex and not easily comparable to mortality studies in the laboratory. A change in population size in experimental plots can be explained not only by increased death rates but also by:

1. Immigration/emigration horizontally or vertically due to escape from treated soil or litter within or between plots (Black & Neely, 1975).
2. Insecticides and fungicides will promote crop growth by suppressing harmful insects or fungal diseases. Herbicides will kill weeds. More crop residues and dead weeds, ie. more food, might compensate for acute toxicity in a long-term study (Boström & Lofs-Holmin, 1986).
3. A change in the microflora may alter decomposition mechanisms making food more or less available (Cook, 1983).
4. Retarded growth and/or reduced reproduction are sublethal effects which may occur in the field. They cannot be measured under field conditions, but these effects will be included in an overall response of a long-term study.

Measured population change is in fact the sum of these uncontrolled factors, which can be artifacts, hidden by indirect effects or overlooked if species composition or age ratio are neglected.

Influence of test design on sensitivity

As so many factors are uncontrolled in the field the sensitivity of the test must be increased by carefully planning experimental design.

1. Experiments should be performed with a randomized block design, with at least four replicates to enable analysis of variance.
2. Soil texture influences sensitivity of the test. Toxicity may be greater in a sandy soil and lower in heavy clay (Lofs-Holmin, 1981).
3. High abundance will make population changes easier to observe. The field can be prepared to ensure high numbers (Boström & Lofs-Holmin, 1986). In arable land total abundance should not be less than 100 individuals per m^2. High abundance of one anecic and one endogeic species should be preferred to high diversity of species.
4. Time of application of a pesticide must be correlated with activity periods of earthworms disregarding whenever it will be used in practice. If earthworms are inactive they may not be exposed to the pesticide and then hazards possibly overlooked.
5. Sampling data should give information about species, individual weights and at least three development stages, ie. adults, sub-adults and juveniles.
6. Sampling dates should be chosen so that both acute toxicity and long-term effects can be detected. For a long-term study one new generation must have hatched, ie. 4–6 months after treatment. Recovery, no statistical decrease in juvenile members compared to control plots after one year, indicates no long-term hazard.
7. Sampling must be efficient and take place when earthworms are active. Experiments should allow all sampling to be done during a single day. Four workers can make 28 extractions per day. Treated plots and controls should be sampled in parallel to compensate for diurnal activity changes. Earthworms can be kept alive in cold water overnight, which will make species determination more reliable also for juveniles.

Discussion

What is a biologically significant effect? Heimbach (1991) suggested a classification so that a reduction in total number of 75% is heavy, 40–75% is medium and 25–40% is weak. Changes less than 25% were not statistically distinguishable.

Application rates must be close to those normally used, for example, maximum recommended and 4× this rate. Higher rates may cause effects which are of no relevance to normal use. On the other hand a normal dose which does not increase mortality may still influence growth and reproduction so that age structure half a year later is altered compared to controls. This stresses the importance of studying age classes. It must also be taken into account that some pesticides will be used annually or several times a year. A population which recovers within a year after a single dose might be more severely affected by repeated treatments.

Tests have to be designed and test parameters chosen with the intention to discover slight effects under worst conditions: 'The worst case' approach. This will provide information on acute toxicity, sublethal effects on population structure and eventual recovery. After test results have been evaluated it must be decided whether a potential hazard can be accepted or not.

A change in earthworm numbers of biomass due to pesticide treatment might not be considered to be of ecological or economical significance as natural variations are much larger and other agricultural practices are more likely to make earthworms disappear, ie. rotary cultivation, burning of crop residues etc. (Lofs-Holmin, 1983). In view of this a 25–30% reduction after one year may not seem to be much, but on the other hand, when put on top of all other hazards also this might be severe and unacceptable.

References

ANDRÉN, O., LINDBERG, T., BOSTRÖM, U., CLARHOLM, M., HANSSON, A-C., JOHANSSON, G., LAGERLÖF, J., PAUSTIAN, K., PERSSON, J., PETTERSSON, R., SCHNÜRER, J., SOHLENIUS, B., WIVSTAD, M. (1990). 5. Organic carbon and nitrogen flows. In: *Ecology of arable land – Organisms, carbon and nitrogen cycling*. Ecological Bulletins, 40. O. Andrén, T. Lindberg, K. Paustian and T. Roswall (Eds.), Copenhagen, pp. 85–126.

BLACK, W., NEELY, D. (1975). Effect of soil-injected benomyl on resident earthworm populations. *Pesticide Science,* **6,** 543–545.

BOSTRÖM, U. (1988). Earthworm populations (*Lumbricidae*) in terminated and undisturbed leys, and their influence on litter disappearance. In: *Ecology of earthworms in arable land. Population dynamics in four cropping systems.* Swedish University of Agricultural Sciences, Department of Ecology and Environmental Research, Dissertation, Report 34, U. Boström, Uppsala, pp 1–17.

BOSTRÖM, U., LOFS-HOLMIN, A. (1986). Influence of fenvalerate, thiabendazole and propiconazole on earthworms in a field experiment. Swedish University of Agricultural Sciences, Department of Ecology and Environmental Research, Subdepartment of Agro-Ecology, Communication 28, Uppsala, pp 45.

BOSTRÖM, U., LOFS-HOLMIN, A. (1988). Earthworm population dynamics and flows of carbon and nitrogen through *Aporrectodea caliginosa (Lumbricidae)* in four cropping systems. In: *Ecology of earthworms in arable land. Population dynamics in four cropping systems*. Swedish University of Agricultural Sciences, Department of Ecology and Environmental Research, Dissertation, Report 34, U. Boström, Uppsala, pp 1–34.

COOK, A. (1983). The effects of fungi on food selection by *L. terrestris*. In: *Earthworm ecology from Darwin to vermiculture*. J.E. Satchell (ed), Chapman and Hall, London, New York, pp. 365–373.

LOFS-HOLMIN, A. (1981). Influence in field experiments of benomyl and carbendazim on earthworms (*Lumbricidae*) in relation to soil texture. *Swedish Journal of Agricultural Research,* **11,** 141–147.

LOFS-HOLMIN, A. (1983). Influence of agricultural practices on earthworms (*Lumbricidae*). *Acta Agriculturae Scandinavica,* **33,** 225–234.

LOFS-HOMIN, A., BOSTRÖM, U. (1988). The use of earthworms and other soil animals in pesticide testing. In: *Earthworms in waste and environmental management*, C.A. Edwards and E.F. Neuhauser (Eds.), SPB Academic Publishing, The Hague, pp 303–313.

Measuring Effects of Pesticides on Earthworms in the Field: Test Design and Sampling Methods

H. KULA

Institute for Zoology, TU Braunschweig, Pockelsstr. 10 a, D-3300 Braunschweig, Germany

Abstract

In order to recognize ecologically significant side effects of pesticides on earthworms, they are often tested in the field. Data from field tests usually show a high variability and interpretation of results is difficult. Therefore a standardized field method for routine testing of pesticides is needed. An experimental design considering current ideas of effect assessment is discussed. Different sampling methods are reviewed concerning their suitability for routine field testing.

Introduction

Boström & Lofs-Holmin (1982) reviewed 63 field studies concerned with the testing of side effects of pesticides on earthworms. A comparative interpretation was difficult because either test design varied between different studies or important information (e.g. soil characteristics) was missing.

In order to standardize experimental design of field tests, a preliminary protocol for field testing is proposed (table 1) and will be discussed.

Experimental design of field tests

SELECTION OF TRIAL SITE

An adequate trial site with sufficient earthworm density and diversity is a prerequisite for reliable test results. There exist at least three possibilities for a field test to take place:

- arable land
- permanent grassland (e.g. pastures, orchards)
- 'artificial' experimental sites (e.g. sown with a grass seed mixture)

Table 1. Preliminary protocol for field testing of pesticides

Trial site:
– grassland
– arable land
Earthworm density:
– minimum density: 100 ind./m^2
– *Lumbricus terrestris* and *Aporrectodea caliginosa* should be present
Plot size:
– minimum size of single plot: 10 x 10 m
Number of replicates:
– at least 4 replicates per treatment
– 4 untreated control plots
– 4 treated plots with highest recommended dose
– 4 treated plots with toxic standard
Sample size:
– depending on the power of study demanded
– at least 2 sample units per replicate (sampling area: 0.25m^2)
Sampling dates:
– treatment in spring
– 1. sampling: 1 month after application
– 2. sampling: 4–6 months after application
– 3. sampling: 12 months after application
Sampling methods:
– formaldehyde method
– electrical octet method

Arable land

This approach meets agricultural practice most closely, but has several disadvantages. First of all, earthworm density and diversity usually are very low. In a survey of 120 bavarian fields, Bauchhenss (pers. comm.) observed low population densities with a median of 8·9 ind./m^2. Patchiness of earthworm distribution seems to be higher than in grassland sites, impeding accurate sampling. *Lumbricus terrestris*, an important earthworm species, is often missing. Agricultural practice like ploughing provides many additional variables which are difficult to consider. If other pesticides have to be applied, the possibility of synergistic effects, as described by Houpert *et al.* (1982), increases and makes it more difficult to relate effects to a single pesticide. If herbicides are applied, the environment (e.g. plant cover of soil surface) in treated plots might differ from control plots.

Permanent grassland

In contrast to arable land, grassland sites usually show high population densities up to 600 ind./m^2 (Edwards & Lofty, 1977). Species diversity is high and all three ecological groups (Bouché, 1977), especially important species of arable land like *Lumbricus terrestris* and *Aporrectodea caliginosa*, are present. Earthworms are more or less homogenously distributed on grassland. The soil structure remains undisturbed by agricultural practices. As other pesticide treatments can be excluded without altering the test system, the test substance can be evaluated alone.

'Artificial' experimental sites

Depending on the grass mixture sown, these sites should have similar characteristics to permanent grassland sites. Given enough time for development, they should sustain sufficient earthworm population densities. Nevertheless, it must be kept in mind that depending on the age of the site different succession stages might be present.

When choosing a trial site the possible range of application of the pesticide has to be considered. Arable land can be used, especially if pesticide effects are expected to depend on application techniques (e.g. in-furrow-application, granules), but the trial sites must have sufficiently high population densities. Because of the disadvantages of arable land, as listed above, in most cases field tests will take place on grassland sites. Effects on important species of the indigenous earthworm fauna can be studied there.

DESCRIPTION OF TRIAL SITE

Biological activity of a pesticide depends on, among others, the chemical properties of the soil (Van Gestel & Ma, 1990). As pointed out by Boström & Lofs-Holmin (1982), information on important soil characteristics of the trial site is necessary for a correct interpretation of the results. Therefore, at least data on soil texture, organic matter content, pH and moisture should be reported.

Information about weather conditions, especially for the date of application, are important for interpreting the results. Photo-oxidation or volatilization of pesticides in dry and sunny periods might considerably reduce the applied amount of pesticide and thus the exposure of earthworms.

GENERAL MAINTENANCE OF FIELD PLOTS

When using grassland, test plots should be mown regularly to maintain a short grass sward. To ensure contamination of the topsoil, test plots should be mown prior to application. The grass cuttings should be left on the ground, as they provide the food source for *Lumbricus terrestris*, but thick grass layers should be avoided as they might prevent soil contamination.

Livestock has to be excluded in order to minimize disturbing variables.

ARTIFICIAL IRRIGATION

Unpredictable dry weather conditions can influence the success of a pesticide application. Therefore it has to be considered whether an artificial irrigation of the test plots can help to standardize soil contamination. In this case the minimum amount of water per m^2 which ensures sufficient infiltration and thus dispersion of the pesticide has to be determined.

Another advantage of artificial irrigation prior to application might be that earthworms will become active and thus exposed to the pesticide. Sampling efficiency might also be improved by irrigation prior to sampling periods.

APPLICATION METHODS AND RATES

Application of pesticides should be done as close as possible to agricultural practice. The highest recommended rate should be applied. For investigation of dose-effect relationships different test plots can be treated with four times the normal rate (Heimbach, pers. comm.). According to common agricultural practice, 300–400 l water/ha should be used for application.

TOXIC STANDARD

The use of a toxic standard is recommended to get information about the level of pesticide effects under the specific conditions at the trial site, the weather, the biological activity, etc. Benomyl, which is known to be very toxic to earthworms (Niklas & Kennel, 1978), can be used for this calibration.

Power of study

PLOT SIZE

The following points have to be considered when fixing plot size. No effects of the sampling whatsoever should occur during the test period. This means that the test plot must be large enough to ensure a minimum distance between adjacent samples. Bouché (1972) proposed a distance of 5 m for population studies, but for routine testing 2 m seems a reasonable compromise. Care must be taken not to sample an area for a second time during successive sampling dates.

As pesticide treatment may represent a disturbing factor for earthworms, besides mortality or inhibition of earthworm activity, hyperactivity and escape reactions might also occur. In a study with a ^{14}C-labelled test substance Shires (pers. comm.) observed radioactivity in earthworms sampled from untreated control plots. The only explanation for this was a postulated migration between treated and untreated plots. In another study (Kula, unpubl.), higher densities of adult *Lumbricus terrestris* were found 4 months after application in insecticide-treated plots when compared to control plots. As reproduction and sexual maturation of *Lumbricus terrestris* takes longer than this, immigration into treated plots may explain these results. Martin (1976) observed more adult *Aporrectodea caliginosa* in pitfall traps during wet periods. He concluded that immigration might contribute considerably to recovery of small test plots. Mather & Christensen (1988) described surface movements of *Lumbricus terrestris*. Mean distance of a single surface foray in one night was $9{\cdot}0 \pm 3{\cdot}8$ m. As the majority of known field studies have used small plot sizes, earthworm migration as natural dispersal behaviour or caused by anthropogenic disturbance might contribute considerably to variability of field data.

Plot sizes in field studies cover the range from small rectangular plots of $1{\cdot}5 \times 5{\cdot}5$ m ($= 8{\cdot}25$ m^2) (Clements *et al.*, 1987) and medium sized quadratic plots of 10×10 m ($= 100$ m^2) (Heimbach, pers. comm.) up to large plots of 50×50 m ($= 2500$ m^2) (McColl, 1984).

To fulfill the requirements stated above, a minimum plot size of 100 m^2 seems reasonable. It should be noticed that this plot size might not exclude bias caused by migration. Whenever possible, bigger plots should be used.

Single plots should be separated by an untreated safety strip with a minimum width of 1–2 m. To avoid edge-effects, samples should be taken only in the inner part of the test plot, such that there is a treated safety strip of at least 1 m.

NUMBERS OF REPLICATES AND SAMPLE SIZE

The number of replicates depends on the level of reliability required. For statistical reasons each treatment including control should be replicated at least four times (Bauchhenss, pers. comm.; Heimbach, pers. comm.).

There is no consensus of the number of sample units per replicate. Bauchhenss (pers. comm.) usually takes 10 sample units, Heimbach (pers. comm.) takes 5 sample units and Edwards & Brown (1982) used only 2 sample units per replicate.

Edwards, P. (pers. comm.) reported that more power is derived from increasing the number of replicates as opposed to the number of sample units when between plot variation is greater than within plot variation. He suggested an experimental design with 2 trial sites and 4 replicates at each site with 2 sample units per sampling date. Data were pooled when trial sites showed similar effects. With these 8 replicates, differences in abundance of 30% could be detected with 90% power but conclusions on species level were not always possible.

Karandinos (1976) published several formulae to calculate optimum sample size. Using these formulae Daniel & Bieri (1988) determined optimum sample sizes for a population of *Lumbricus terrestris* to be 8–12 sample units for adults and 3 sample units for juveniles.

As the optimum sample size depends on density and distribution of earthworm populations, it has to be calculated for the specific circumstances of the trial site under investigation. Especially when interpreting results on species level, low mean values might require higher numbers of sample units. Multiple samples per experimental unit increase the sensitivity of the test by increasing the precision with which mean abundance of each replicate is estimated.

DURATION OF THE STUDY

Test duration depends on the aims of the study as well as on characteristics of the test substance. For routine testing it seems convenient to take an investigation period of about one year. In agricultural practice the pesticide might then be applied again, therefore earthworm populations should at least have recovered in this time. In cases of very persistent pesticides, or if bioaccumulation in food-chains or other long-term effects are to be expected, longer investigation periods should be considered.

SAMPLING DATES

Because of the long generation time for earthworms (Satchell, 1967; Edwards & Lofty, 1977), the time between application and the first and last sampling date is important for recognizing all significant effects. Short-term acute effects should be considered as well as sublethal long-term effects, especially on reproduction. Therefore sampling dates should cover the activity periods in spring and in autumn. As a prerequisite when using formaldehyde or electrical sampling methods (cf. behavioural sampling methods) it has to be ensured that species of all ecological groups are in an active phase when sampling takes place. A widely accepted sampling pattern (Edwards & Brown, 1982; Heimbach, 1990; Bauchhenss, pers. comm.) proposes 3 sampling dates. Very often application will take place in spring or early summer. To observe short-term acute effects, first samples should be taken about 1 month later. As some earthworm species aestivate during the summer, the next samples should be taken 4–6 months after application in autumn which is the main period of earthworm activity. Sublethal effects on ecologically significant functions like reproduction and hatching of juveniles can then be estimated. The last samples should be taken 12 months after application to determine long-term effects and especially recovery.

Sampling methods

GENERAL REMARKS

Earthworm sampling methods were reviewed by Satchell (1969), Nordström & Rundgren (1972), Bouché & Gardner (1984) and Lee (1985). Quantitative sampling of abundance and biomass is very difficult, because earthworm distribution in the soil is not homogenous (Guild, 1952). Depending on their ecological strategy, earthworms are active at different depths of the soil. Activity can vary between species, between life-stages within species, and also vary seasonally. Diurnal activity changes which cause extraction efficiency changes have been described for *Lumbricus terrestris* (Daniel, 1986). Besides these variations due to the biology of earthworms, sampling methods also depend on characteristics of the trial site, such as physical properties of the soil, soil moisture and soil temperature. Therefore unit sampling areas and sample depths must be fitted to the density and distribution of earthworm populations under investigation. All methods of collection and extraction have their own drawbacks and therefore are under-estimates of populations.

Lee (1985) subdivided the different sampling methods into two main categories:

– passive methods (e.g. handsorting, washing and sieving, flotation)
– behavioural methods (e.g. formaldehyde expulsion, electrical sampling)

PASSIVE METHODS

Earthworms are physically sorted from soil samples by means of handsorting, washing and sieving or flotation. These methods give the most accurate estimates but are very laborious. As all of these methods depend on samples isolated from the soil, estimates are limited to the volume and depth of the soil core. Adult individuals of deep-burrowing species may, therefore, be underestimated or not detected. Handsorting may also underestimate small species and especially immatures of all species. It is assumed that specimens < 200 mg live weight or < 2 cm length are generally overlooked (Lee, 1985). Overall efficiency ranges between 52% to 93% of numbers and 84% to 99% of biomass of a population but species-specific differences are observed (Lee, 1985).

Results can be optimized by a combination of handsorting with washing and sieving techniques. Walther & Snider (1984) reported efficiencies of 98% of numbers and nearly 100% of biomass for humus and soil samples.

BEHAVIOURAL METHODS

Formaldehyde method

Among the chemicals which have been used to expell earthworms from the soil, formaldehyde solution (Raw, 1959) gives the most satisfactory results. The method depends on the activity of earthworms, therefore inactive, aestivating or senescent individuals are not detected. It is not quite clear whether a relation between the number of collected earthworms and the soil surface area or the volume of the sample can be established. As the formaldehyde solution will predominantly infiltrate into the vertical burrows of anecic species, earthworms which live in more horizontal burrow systems like most mineral soil dwellers will be underestimated. Raw (1959) found the method most effective for *Lumbricus terrestris*, but efficiencies for mineral soil dwellers were only 60% for *Allolobophora chlorotica*, 61% for *Aporrectodea caliginosa*, 45% for *Aporrectodea longa* and 57% for the epigeic *Lumbricus castaneus*. Compared to handsorting, Nordström & Rundgren (1972) found significantly fewer *Aporrectodea rosea* and *Aporrectodea caliginosa*. In *Aporrectodea caliginosa* this difference was due to juvenile life-stages only because no significant differences were found in adults and subadults.

Electrical sampling

The electrical octet-method developed by Thielemann (1986) offers a non-destructive earthworm sampling method, which is particularly useful for study on field plots. In contrast to other methods, where the plant-soil system is destroyed by digging and successive handsorting or by saturation of the soil with chemical expellants, disturbance is negligible because only 8 electrodes have to be pushed into the soil. Thielemann (1986) reported mean extraction efficiencies of 87·7% of numbers. Kula (in prep.), in contrast, found lower mean efficiencies of 66% of numbers and 65% of biomass. Extraction efficiency varied between species. From biomass data it could be seen that adult and subadult *Lumbricus terrestris* were underestimated, whereas estimates of the abundance of juvenile life-stages were similar to results with formaldehyde. Extraction efficiency for endogeic species varied with soil moisture, and especially *Aporrectodea rosea* was poorly represented. In order to improve efficiency, Cuendet *et al.* (1991) developed longer, segmented electrodes and a sophisticated extraction programme. This modification especially improved extraction of adult *Lumbricus terrestris*. Compared to the formaldehyde method they did not find significant differences in extraction efficiency of common anecic, endogeic and epigeic earthworm species.

A major limitation of the electrical sampling method is soil moisture because certain soil humidity is necessary to establish the electrical field. Other inhomogenities like stones in the soil etc. may also influence efficiency.

Biomass estimation

In many field studies, only data on earthworm abundance are analysed, but important information on sublethal effects of pesticides, such as growth inhibition of juveniles, can also be derived from biomass data. Lofs-Holmin (1982) observed reduced body weights of juvenile *Aporrectodea caliginosa* and *Lumbricus terrestris* in treated plots while earthworm abundance was not affected.

Correct biomass estimation is difficult. Fully hydrated earthworms contain about 80% of water, but depending on activity level and soil moisture they are able to loose considerable amounts of water. Therefore biomass estimates should be expressed in terms of dry weight (Lee, 1985). The gut contents, which can comprise a considerable amount of the live weight, should be excluded because it varies with feeding and burrowing behaviour (Lee, 1985).

For reasons stated above, in routine field testing only the weight of preserved specimens will be determined. If data on life weight or dry weight are needed for special purposes, conversion factors for most species are available in literature.

METHODS RECOMMENDED FOR ROUTINE FIELD TESTS

In routine field tests many plots have to be sampled in very short time to allow comparisons. Only methods which demand little effort can therefore be considered.

Though very accurate in estimating abundance and biomass of earthworms as well as cocoons, all passive methods therefore usually will be disregarded because they are too laborious. The use of these methods will be confined to comparative population studies in different ecosystems.

For the comparison of treated and untreated plots, relative estimates of abundance and biomass of earthworms seem to be sufficient, when the extraction efficiency is similar for all species or possible differences are taken into account.

Bouché & Gardner (1984) stated that the formaldehyde method, though always underestimating actual numbers, can give comparable results when samples are all taken at the same time from soil with the same soil moisture and temperature. As control and treated plots should be situated closely together on a more or less homogenous trial site these conditions might be realized. Use of the formaldehyde method therefore can be recommended.

Although still in a stage of improvement the electrical octet method can also be used, especially when disturbance or destruction of sample areas is not desired.

Routine checking of extraction efficiency is strongly recommended for both behavioural methods.

Outlook

Is it realistic to assume that all biologically significant effects, especially long-term effects, can be recognized by only one field test according to the experimental design described above? From the ecological point of view, Jepson (1988) demanded more than one test, with sufficient attention paid to the recovery of populations. This seems reasonable because the results of a single test might depend on the specific spatial and temporal conditions during the test (e.g. the importance of application time in relation to activity of earthworms). Extrapolation from the results then is questionable. Another point is the threshold for an effect to be considered critical from the ecological point of view. Further research is needed to answer these questions.

References

BOSTRÖM, U., LOFS-HOLMIN, A. (1982) Testing side effects of pesticides on the soil fauna – a critical literature review. *Report 12, Swedish University of Agricultural Sciences, Dep. of Ecology and Environmental Research*, Uppsala.

BOUCHÉ, M. (1972) Lombriciens de France. Ecologie et Systematique. *INRA Publ. 72–2*, Paris.

BOUCHÉ, M. (1977) Strategies lombriciennes. *Ecological Bulletin,* **25,** 122–132.

BOUCHÉ, M., GARDNER, R. (1984) Earthworm functions 8. Population estimation techniques. *Revue d' Ecologie et Biologie du Sol,* **21,** 37–64.

CLEMENTS, R.O., BENTLEY, B.R., NUTTAL, R.M. (1987) The invertebrate population and response to pesticide treatment of two permanent and two temporary pastures. *Annals of Applied Biology,* **111,** 399–407.

CUENDET, G., BIERI, M., VETTER, F. (1991): Elektrische Regenwurmextraktions methode. *Schriftenreihe Umwelt*, Nr. 154, *Bundesamt für Umwelt, Wald und Landschaft*, Bern.

DANIEL, O. (1986) Diurnale Effizienzschwankungen beim Sammeln der Regenwurmart *Lumbricus terrestris* mit der Formaldehydmethode. *Bulletin Bodenkundliche Gesellschaft der Schweiz,* **10,** 39–44.

DANIEL, O., BIERI, M. (1988) Stichprobenplan für die Regenwurmart *Lumbricus terrestris L. Bulletin Bodenkundliche Gesellschaft der Schweiz,* **12,** 195–198.

EDWARDS, C.A., LOFTY, J.R. (1977) *Biology of earthworms.* 2nd Ed., London: Chapman and Hall.

EDWARDS, P.J., BROWN, S.M. (1982) Use of grassland plots to study the effect of pesticides on earthworms. *Pedobiologia,* **24,** 145–150.

GUILD, W.J.MCL. (1952) Variation in earthworm numbers within field populations. *Journal of Animal Ecology,* **21,** 169–181.

HEIMBACH, F. (1990): Beeinflussung der Regenwurmfauna einer Graslandfläche durch Schneckenkorn Mesurol. *Pflanzenschutz-Nachrichten Bayer,* **43,** 140–150.

HOUPERT, G., JENOT, M., LARDIER, P.A. (1982) La sensibilité accrue d' *Eisenia fetida* (Lumbricidae) aux insecticides carbamates en presence d' atrazine. *Bulletin E.N.S.A.I.A,* **24,** 3–9.

JEPSON, P.C. (1988) Ecological characteristics and the susceptibility of non-target invertebrates to long-term pesticide side effects. In: *Field methods for the study of environmental effects of pesticides*, Greaves, M.P.; Smith, B.D.; Greig-Smith, P.W. (Eds.), *BCPC Monograph No. 40,* Thorton Heath: BCPC Publications, pp. 191–200.

KARANDINOS, M.G. (1976) Optimum sample size and comments on some published formulae. *Bulletin of the Entomological Society of America,* **22,** 417–421.

LEE, K.E. (1985) *Earthworms, their ecology and relationships with soils and land use.* Sydney: Academic Press, 411 pp.

LOFS-HOLMIN, A. (1982) Influence of routine pesticide spraying on earthworms (Lumbricidae) in field experiments with winter wheat. *Swedish Journal of Agricultural Research,* **12,** 121–123.

MARTIN, N.A. (1976) Effect of four insecticides on the pasture ecosystem. Part V: Earthworms (Oligochaeta: Lumbricidae) and Arthropoda extracted by wet-sieving and salt flotation. *New Zealand Journal of Agricultural Research,* **19,** 111–115.

MATHER, J.G., CHRISTENSEN, O. (1988) Surface movements of earthworms in agricultural land. *Pedobiologia,* **32,** 399–405.

MCCOLL, H.P. (1984) Nematicides and field populations of enchytraeids and earthworms. *Soil Biology and Biochemistry,* **16,** 139–143.

NIKLAS, J., KENNEL, W. (1978) Lumbricidenpopulationen in Obstanlagen der Bundesrepublik Deutschland und ihre Beeinflussung durch Fungizide auf Basis von Kupferverbindungen und Benzimidazolderivaten. *Zeitschrift für Pflanzenkrankheiten und Pflanzenschutz,* **85,** 705–713.

NORDSTRÖM, S., RUNDGREN, S. (1972) Methods of sampling lumbricids. *Oikos,* **23,** 344–352.

RAW, F. (1959) Estimating earthworm populations by using formalin. *Nature,* **184,** 1661–1662.

SATCHELL, J.E. (1967) Lumbricidae. In: *Soil Biology*, Burges, A. & Raw, F. (Eds.), London: Academic Press, pp. 259–322.

SATCHELL, J.E. (1969) Methods of sampling earthworm populations. *Pedobiologia,* **9,** 20–25.

THIELEMANN, U. (1986) Elektrischer Regenwurmfang mit der Oktett-Methode. *Pedobiologia,* **29,** 297–303.

VAN GESTEL, C.A.M., MA, W. (1990) An approach to quantitative structure-activity relationsships (QSAR's) in earthworm toxicity studies. *Chemosphere,* **21,** 1023–1033.

WALTER, P.B., SNIDER, R.M. (1984) Techniques for sampling earthworms and cocoons from leaf litter, humus and soil. *Pedobiologia,* **27,** 293–297.

Effects of Pesticides on Earthworm Populations: Comparison of Results from Laboratory and Field Tests

F. HEIMBACH

Bayer AG, Agrochemicals Research, Environmental Biology, 5090 Leverkusen, Germany

Abstract

A laboratory procedure using an artificial soil ('Artificial Soil Test') and *Eisenia fetida* as test species is widely used for testing the toxicity of pesticides to earthworms. To validate the results of this test in respect to natural earthworm populations, and to allow the ecological significance of the laboratory data to be interpreted and extrapolated to the field, a field test on perennial grassland was developed. 21 field studies using 12 different pesticides were conducted; the biomass and abundance of earthworms shortly after application (4–6 weeks) of the highest recommended application rate and the four-fold overdose of pesticide, in the autumn of the same year and the spring of the following year were determined. An exposure calculation was proposed which allowed comparison of laboratory LC_{50} values to 'estimated environmental concentrations' (EEC) in soils derived from field application rates. Several similar long term studies of other authors testing some pesticides on grassland areas have been evaluated. A good correlation was found between laboratory results and effects on earthworm populations in the field (r = 0·86, n = 29).

Introduction

A laboratory procedure, the 'Artificial Soil Test', which was developed to determine the toxicity of chemicals to earthworms (EEC, 1985; OECD, 1984), has been tested by several authors (Callahan *et al.*, 1989; Edwards, 1984; V. Gestel & V. Dis, 1988; Heimbach, 1984, 1988) and now appears to be widely accepted as a standard test for pesticides. Since the use of the 'Artificial Soil Test' as a standard laboratory test can only be justified if results from laboratory and field experiments are sufficiently correlated, V. Gestel (in press) and Heimbach (1992) compared data from laboratory and field experiments. In the present paper, data from Heimbach's (1992) work are compared to results of other authors who have conducted similar long-term tests.

Materials and methods

The methods used have been described and discussed in detail by Heimbach (1992). Both laboratory and field tests were conducted with 21 products based on 12 active ingredients: azinphos-methyl (EC-formulation), benomyl (WP), captafol (SK), captan (WP), cyfluthrin (EC), endosulfan (EC), ethiofencarb (EC), fenamiphos (EC), imidacloprid (proposed common name) (WP), mercaptodimethur (WP), oxydemeton-methyl (EC) and propoxur (EC). The products have been numbered and listed in Table 1. Nos. 1, 4, 12, 13 and 18 are formulations of three fungicidal compounds, all others are insecticides.

In the 'Artificial Soil Test', earthworms are exposed to different concentrations of pesticide in a substrate composed of peat, sand and clay mineral. The animal used is *Eisenia fetida andrei* (Bouché, 1972). This earthworm can readily be bred, and its sensitivity to pesticides is comparable to that of *Lumbricus terrestris* (Haque & Ebing, 1983; Heimbach, 1985). The duration of the 'Artificial Soil Test' is 14 days. The data determined are the LC_{50} (concentration at which 50% of the earthworms die), the highest concentration without observed effects (NOEC), and the lowest concentration with effects (LOEC). Because of the sigmoidal dose-activity relationship and the spacing of the concentrations used in the experiment, the LC_{50} appears to be the most suitable value for the assessment of the toxicity. This value can be determined with the highest degree of accuracy and can be statistically calculated with the 95% confidence interval. For this reason, the LC_{50} values will be generally referred to.

The standardized field experiments were conducted from 1980 to 1988 on pastures near Leverkusen (Germany). The pastures had a high abundance (up to 800 individuals per m^2) and a high diversity (up to 8 species) of earthworms. Plots of 10 × 10 m were treated in duplicate with the highest registered application rate of each pesticide, and with four times this rate. Following the directions for use, some pesticides were applied several times (Table 1). The earthworm abundance was determined 4 to 6 weeks after the first application in May of each year, thereafter in the autumn, and again in the spring of the following year. Five samples were evaluated per plot, i.e. 10 samples per product, rate and date. The results from treated plots were compared to those from two untreated control plots.

Results

Table 1 compares the results from laboratory and field tests with 21 formulations of 12 active ingredients. The products were selected on the basis of data from preliminary laboratory tests. Pesticides with LC_{50} values above 600 mg/kg were not examined in the field since toxic effects to earthworm populations could be excluded even at high application rates.

In laboratory experiments LC_{50} values were determined by using *Eisenia fetida*. As suggested by Heimbach (1992) in the procedure for comparison of laboratory and field results, the LC_{50} value is divided by a compensation factor of 10 before it is compared to the 'estimated environmental concentration' (EEC). This factor is large enough to cover the different sensitivities of various earthworm species (which is generally sufficiently taken into consideration using a factor of 2–3; Haque & Ebing, 1983; Heimbach, 1985). It is generally also large enough to account for flat dose responses. Nevertheless, certain NOEC values from the laboratory test should be taken into consideration: if the NOEC/LOEC range is much lower than the LC_{50}, e.g. by the factor of 40 (= very flat dose-activity curve), the values must be interpreted with care.

Beside the LC_{50}'s from the laboratory tests, Table 1 shows the respective rates of application and their effects on the earthworm fauna: In this case, a reduction of the earthworm abundance of more than 75% was classified as heavy, about 40–75% as medium, and about 25 to 40% as weak. Changes in earthworm abundance of less than 25% are, in most cases, not statistically distinguishable because of the variable distribution of earthworms. Changes in abundance classified as heavy and medium refer to total earthworms (all species) while weak effects were generally restricted to 1 or 2 species.

Discussion

Exposure of earthworms to pesticides in the field has to be estimated, based on both the distribution of the chemical and earthworms in the soil. In this paper estimation of the chemical concentration was based on a uniform distribution of the pesticide in the top 2·5 cm of soil, and an average soil density of 1·5 kg/l. Accordingly, 1 kg active ingredient applied to 1 ha resulted in an 'estimated environmental concentration' (EEC) of 2·7 mg AI/kg dry soil. Depending upon the properties of both the pesticide and the soil, as well as on weather conditions and mode of application, this assumption certainly does not always apply. However, in our experience, it offers a reasonable value for the assessment of a potential hazard. In addition, concentration calculations and analytical results for several pesticides in natural soils by some authors justify this proposal (Beyer, this volume; Gerber *et al.*, 1989; van Gestel, in press; Korschgen, 1970; Lichtenstein *et al.*, 1960).

Some pesticides must be applied several times during a growing season to achieve a sufficiently high degree of effectiveness. This is particularly true for some insecticides and fungicides, especially if these are rapidly degradable. The pesticides used in our studies are degradable under the climatic conditions in the field tests. Thus, it would be incorrect to calculate the final concentrations in the soil as simple multiples of the individual treatments per season. As a first approach, the use pattern is therefore included in a very simplified and practicable procedure which does not indicate real concentrations: to the full amount of the first application (i.e. 100%), 50% of each further application is added (resulting e.g. for three equal applications at $100 + (2 \times 50) = 200\%$ of the individual applications). This is obviously a very 'conservative' estimation, the real maximum concentrations in the field will certainly be considerably lower.

The concentrations used in the various studies (Tables 1 and 2) were calculated as described above. The resulting concentrations are compared with the laboratory results, the 'reference value' being 1/10 of the LC_{50} in % of the field EEC. Thus, a value of 100% shows equality of the concentration in the field and the compensated laboratory value. Lower values indicate that the concentrations in the field were higher than the resulting values in the laboratory, higher values indicate that the concentrations in the field are lower than the corresponding laboratory concentrations (i.e. 1/10 of the LC_{50}). The results of our own studies (Table 1) show that the LC_{50} value allowed a fair prediction of the effects of the pesticides on field populations of earthworms. If the 'reference value' was below 30–40%, a reduction of the field populations of earthworms was observed in all cases. Long-term reductions were only observed when the reference values were below 10. In all other cases, hardly any or only transient effects were found.

Table 2 summarizes data obtained by other authors for benomyl and the insecticides carbofuran and carbaryl on grassland areas. Only long term studies using a test design similar to that described above were used for comparison of standardized laboratory and field tests. The results in Table 2 are in good agreement with those in Table 1. Benomyl, a fungicide known

Table 1. Toxicity of pesticides to earthworms: Comparison of laboratory and field data (data from Heimbach, 1992).

No	Laboratory LC_{50} (mg AI/kg dry soil)	% AI in formul.	Field Experiments: application rate (kg formul./ha)	EEC[1] (mg/kg)	results after[2]: 4–6 weeks	½ year	1 year	Reference value: $LC_{50}/10$ in % of EEC
1	19	50	2 × 16	32	***	***	**	5·9
2	9·4	35	3 × 6	11	–[3]	***	***	8·4
3	11	25	2 × 8	8·0	***	**	*	14
4	19	50	2 × 4	8·0	***	**	*	24
5	59	20	2 × 30	24	***	**	0	25
6	318	40	1 × 100	106	**	**	0	30
7	27	20	5 × 4	6·4	–[3]	***	*	42
8	11	25	2 × 2	2·0	**	0	0	55
9	129	50	2 × 8	16	*	0	0	81
10	59	20	2 × 7·5	6·0	0	0	0	98
11	318	40	1 × 25	27	*	*	0	119
12	516	48	2 × 18	35	0	0	0	149
13	612	50	12 × 4	35	–[3]	0	0	177
14	262	50	2 × 4·8	9·6	0	0	0	273
15	115	25	2 × 4	4·0	0	0	0	287
16	129	50	2 × 2	4·0	0	0	0	322
17	15	5	2 × 2	0·4	*	0	0	375
18	358	48	2 × 4·5	8·7	0	0	0	414
19	262	50	2 × 1·2	2·4	0	0	0	1092
20	115	25	2 × 1	1·0	0	0	0	1150
21	15	5	2 × 0·5	0·1	0	0	0	1500

[1] EEC = 'estimated environmental concentration' in soil (100% of first and 50% of subsequent applications summed up and referred to top 2·5 cm of soil)
[2] ***, **, *, 0 = heavy, medium, weak, no effects on earthworm populations
[3] no samples taken, plot size reduced

Table 2. Toxicity of pesticides to earthworms: Comparison of laboratory and field data found by several authors in standardized studies.
The laboratory LC_{50}'s (*Eisenia fetida*) for these pesticides are: Benomyl 19 mg/kg (Table 1, No. 1 and 4), Carbofuran 5–10 mg/kg (Heimbach 1985); Carbaryl 69 mg/kg (Ammon 1985) – 174 mg/kg (Heimbach 1984).

Pesticide (Reference)	formul.	application rate (kg AI/ha)	EEC[1] (mg/kg)	results after[2] 3–5 weeks	results after[2] ½ year	results after[2] 1 year	Reference value: LC_{50}/10 in % of EEC
Benomyl							
	WP	1 × 10	27	***	***	***	7.1
(Edwards &		1 × 5	13	***	**	*	14
Brown 1982)		1 × 2	5.3	**	*	0	36
(Tomlin *et*	WP	1 × 2.2[5]	6.0	**	**	**	32[5]
al. 1980/81)		1 × 1·1[5]	3.0	*	**	**	64[5]
Carbofuran							
(Thompson & Sans 1974)	WP	1 × 4.5	12	***	–[3]	0[4]	4.2–8.4
(Martin 1976)	WP	1 × 2·2	6·0	**	*	–	8.4–17
Carbaryl							
(Thompson & Sans 1974)	WP	1 × 2·2	6.0	*	–[3]	0[4]	116–291

[1], [2] see Table 1
[3] no samples taken
[4] plot size very small
[5] The application rate of 0·56 kg AI/ha resulted to same effects as the application rate of 1·12 kg AI/ha in 1974, but a second application of 0·56 hg AI/ha in 1975 did not indicate any effect on earthworm populations. As these studies have been performed on large areas along airport runways instead of replicated plots, results have to be compared to untreated control areas with care.

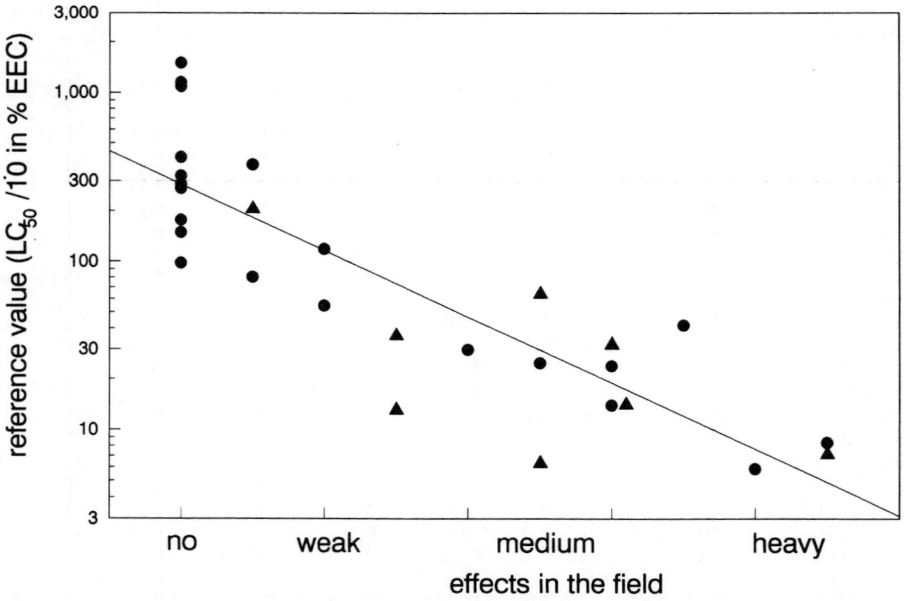

Figure 1 Correlation between the 'reference value' and biological effects of different pesticides in the field. (circles = results of Heimbach, 1992, Table 1; triangles = results of other authors, Table 2)

to be toxic to earthworms (Cook & Swait, 1975; Edwards & Brown, 1982; Stringer & Lyons, 1974, 1977; Stringer & Wright, 1973) was found to be very toxic to field populations of earthworms at high application rates, and moderately toxic at lower rates. The comparison of laboratory and field toxicities for earthworms (the 'reference values') are also very similar to those in Table 1. Figure 1 compares the 'reference values' to the observed effects of the pesticides in the field. To prepare this figure, effects observed a few weeks, 6 months and 1 year after application were summarized: heavy effects were given 3 points, medium effects 2 points, weak effects 1 point, and no effects 0 points. In cases where data were not available, results were extrapolated.

A good correlation was obtained between LC_{50} values from the 'Artificial Soil Test' and a standardized field test conducted on grassland plots. The correlation coefficient, $r = 0{\cdot}86$, indicates a significant correlation of this data set ($\ln y = 5{\cdot}66 - 0{\cdot}45\ x$, $n = 29$). Hence, this test fulfils the requirements of a laboratory test for hazard assessment. On the other hand, the standardized field test using grassland plots allows a ranking of toxic effects of pesticides on earthworm populations.

References

AMMON, H.U. (1985) Worm toxicity tests using *Tubifex tubifex*. In: *Comportment et effets secondaires des pesticides dans le sol*, Versailles 4–8 Juin 1984. Ed INRA Publ. (Les Colloques de l'INRA, No. 31), pp. 303–317

BEYER, W.N. (1992) Relating results from earthworm toxicity tests to agricultural soil. In *Ecotoxicology of Earthworms* (H. Becker *et al.* eds) pp. 109–115 Intercept Ltd, Hants, UK.

BOUCHÉ, M.B. (1972) Lombriciens de France, Écologie et Systématique. Publ. Institut National de la Recherche Agronomique, Paris.

CALLAHAN, C.A., WILBORN, D., MORCOCK, R.E., NEUHAUSER, E.F. AND DREWES, C.D. (1989) Earthworms in terrestrial toxicity assessment. 10th Annual Meeting, Society of Environmental Toxicology and Chemistry (SETAC), Toronto, Canada, October/November 1989

COOK, M.E. AND SWAIT, A.A.J. (1975) Effect of some fungicide treatments on earthworm population and leaf removal in apple orchards. *The Journal of Horticultural Science* **50** (4), 495–499

EDWARDS, C.A. (1984) Report of the second stage in development of a standardized laboratory method for assessing the toxicity of chemical substances to earthworms. Commission of the European Communities, Brussels

EDWARDS, P.J. AND BROWN, S.M. (1982) Use of grassland plots to study the effect of pesticides on earthworms. *Pedobiologia* **24,** 145–150

EEC (European Economic Community) (1985) Toxicity for earthworms – Artificial soil test. EEC Directive 79/831, Annex V, Part C: Methods for the determination of ecotoxicity – Level 1. June 14, 1985

GERBER, H.R., ANDERSON, J.P.E., BÜGEL-MOGENSEN, B., CASTLE, D., DOMSCH, K.H., MALKOMES, H.-P., SOMERVILLE, L., ARNOLD, D.J., VAN DE WERF, H., VERBEKEN, R. AND VONK, J.W. (1989) 1989 Revision of recommended laboratory tests for assessing side-effects of pesticides on soil microflora. 4th International Workshop in Basle, Switzerland, September 1989. Publ.: Bundesforschungsanstalt für Landwirtschaft, Braunschweig (Germany)

VAN GESTEL, C.A.M. AND VAN DIS, W.A. (1988) The influence of soil characteristics on the toxicity of four chemicals to the earthworm *Eisenia fetida andrei* (Oligochaeta). *Biology and Fertility of Soils* **6,** 262–265

VAN GESTEL, C.A.M. (in press) Validation of earthworm toxicity tests by comparison with field studies: A review on benomyl, carbendazim, carbofuran and carbaryl.

HAQUE, A. AND EBING, W. (1983) Toxicity determination of pesticides to earthworms in the soil substrate. *Zeitschrift für Pflanzenkrankheiten und Pflanzenschutz* **90,** 395–408

HEIMBACH, F. (1984) Correlation between three methods for determining the toxicity of chemicals to earthworms. *Pesticide Science* **15,** 605–611

HEIMBACH, F. (1985) Comparison of laboratory methods, using *Eisenia foetida* and *Lumbricus terrestris*, for assessment of the hazard of chemicals to earthworms. *Zeitschrift für Pflanzenkrankheiten und Pflanzenschutz* **92** (2), 186–193

HEIMBACH, F. (1988) A comparison of laboratory methods for toxicity testing with earthworms. In: *Earthworms in waste and environmental management,* Ed.: C.A. Edwards & E.F. Neuhauser, SPB Academic Publishing, The Haque, pp. 329–335

HEIMBACH, F. (1992) Correlation between data from laboratory and field tests for investigating the toxicity of pesticides to earthworms. Proceedings International Symposium on Earthworm Ecology, Avignon (France), June 1990· *Soil Biology and Biochemistry* (in press).

KORSCHGEN, L.J. (1970) Soil-food-chain-pesticide wildlife relationships in Aldrin-treated fields. *Journal of Wildlife Management* **34** (1), 186–199

LICHTENSTEIN, E.P., DE PEW, L.J., ESHBAUGH, E.L. AND SLEESMAN, J.P. (1960) Persistence of DDT, Aldrin, and Lindane in some Midwestern soils. *Journal of Economic Entomology* **53** (1), 136–142

MARTIN, N.A. (1976) Effect of four insecticides on the pasture ecosystem. V. Earthworms (Oligochaeta: Lumbricidae) and Arthropoda extracted by wet sieving and salt flotation. *New Zealand Journal of Agricultural Research* **19,** 111–115

OECD (Organization for Economic Cooperation and Development) (1984) Earthworm, acute toxicity test. OECD Guideline for testing of chemicals, No. 207. April 4, 1984

STRINGER, A. AND WRIGHT, M.A. (1973) The effect of benomyl and some related compounds on Lumbricus terrestris and other earthworms. *Pesticide Science* **4,** 165–170

STRINGER, A. AND LYONS, C.H. (1974) The effct of benomyl and thiophanate methyl on earthworm populations in apple orchards. *Pesticide Science* **5,** 189–196

STRINGER, A. AND LYONS, C.H. (1977) The effect on earthworm populations of methods of spraying benomyl in an apple orchard. *Pesticide Science* **8,** 647–650

THOMPSON, A.R. AND SANS, W.W. (1974) Effects of soil insecticides in Southwestern Ontario on non-target invertebrates: Earthworms in pasture. *Environmental Entomology* **3,** 305–308

TOMLIN, A.D., TOLMAN, J.H. AND THORN, G.D. (1980/81) Suppression of earthworm (*Lumbricus terrestris*) populations around an airport by soil application of the fungicide, benomyl. *Protection Ecology* **2,** 319–323

Estimation of chemical exposure to earthworms

Relating Results from Earthworm Toxicity Tests to Agricultural Soil

W.N. BEYER

Patuxent Wildlife Research Center, US Fish and Wildlife Service, Laurel, MD 20708, USA

Abstract

The artificial soil tests of the European Economic Community and of the Organization for Economic Cooperation produce data relating earthworm mortality to pesticide concentrations in soil under laboratory conditions. To apply these results to agricultural soils it is necessary to relate these concentrations to amounts of pesticide applied per area. This paper reviews the relevant published literature and suggests a simple relation for regulatory use. Hazards to earthworms from pesticides are suggested to be greatest soon after application, when the pesticides may be concentrated in a soil layer a few millimeters thick. For estimating exposure of earthworms, however, a thicker soil layer should be considered, to account for their movement through soil. During favorable weather conditions, earthworms belonging to species appropriate to the artificial soil test have been reported to confine their activity to a layer about 5 cm. If a 5-cm layer is accepted as relevant for regulatory purposes, then an application of 1 kg/ha would be equivalent to 1·67 ppm (dry) in the artificial soil test.

Introduction

Earthworm toxicity tests, such as the artificial soil tests of the European Economic Community (1985) and of the Organization for Economic Cooperation and Development (1984), relate pesticide concentrations in soil to toxic effects on earthworms. When the purpose of such a test is to predict effects on earthworms under natural agricultural conditions, then toxic soil concentrations, expressed as pesticide per weight of soil, must be related to the amount of pesticide per area that might be applied. Without knowing this relation, it is difficult to use the toxicity data in regulatory procedures, other than to rank the pesticides by relative toxicity. Several authors have related the two measures of pesticides in their studies, but none of the relations have become generally accepted. The need for such a relation has been mentioned by Van Leemput *et al.* (1989), and Fred Heimbach has identified this as an important topic for the International Workshop on Earthworm Toxicology.

Although toxicity tests could be designed to simulate local conditions, most of them, including the artificial soil test, standardize the conditions. The procedure for the artificial soil test is the same, no matter what particular soil conditions and earthworm species might be expected when the pesticide is to be applied. To be consistent with this approach, and to make applying the results of the earthworm toxicity test practical, it seems advisable to convert from ppm to kg/ha with a constant factor, rather than use an approach that varies with the pesticide and application conditions. This approach is not meant to imply that the relation is not site specific, but rather that the advantages of maintaining standardization outweigh the disadvantages associated with trying to model what is a very complex and sometimes rapidly changing relation.

This paper first mentions the approaches used in related toxicological evaluations. Then three examples are given, in which researchers have converted toxic soil concentrations from laboratory studies on earthworms to kg/ha. The paper reviews how researchers have sampled soils for pesticides when studying the uptake of pesticides by earthworms, and then discusses the distribution of pesticides and the movement of earthworms in soil. Finally, for the purposes of regulation, the paper recommends assuming that a pesticide becomes distributed to a depth of 5 cm soon after application, and argues that this soil layer is relevant to consider for evaluating hazards to earthworms.

Discussion

PROCEDURES USED IN RELATED EVALUATIONS

The artificial soil test has much in common with laboratory tests measuring the effects of pesticides on soil microorganisms. Heimbach (1991) reports that both 5 cm and 10 cm have been recommended as the appropriate depth for applying laboratory results on soil microorganisms to field conditions. In evaluating phytotoxic hazards from food or feed additives or from animal drugs in waste or sewage sludge when applied to farms, the U.S. Food and Drug Administration assumes that the material will be mixed to a depth of 15 cm (Harrass *et al.*, 1991). The pesticide registration procedure of the U.S. Environmental Protection Agency (1986) for evaluating hazards to wildlife does not involve soil concentrations. Instead it assumes that hazard is greatest soon after application, and from a large data base available (Hoerger and Kenaga, 1972) relates the amount of pesticide applied per area to predicted residues on vegetation. The vegetation is considered potential food for wildlife and the predicted residues in the vegetation are compared to toxicity data on wildlife. Each of these three evaluation procedures converts toxicity data to amounts applied per area. The three procedures are extremely general and do not account for chemical differences in the toxic substances applied or site-specific factors.

CONVERSION FROM PPM IN LABORATORY STUDIES TO KG/HA

Van Leemput *et al.* (1989) measured the toxicity of Imazalil to earthworms in a laboratory study and then, by assuming that this fungicide would be distributed in a 15-cm layer in the field, estimated that 0·5 ppm in the experimental soil would be equivalent to 1 kg/ha. Pižl (1988) used a conversion factor of 1·8 ppm equaling 1 kg/ha to evaluate laboratory results, but did not explain the reasoning behind this factor. Heimbach (1991) mentioned the need to protect earthworms living close to the surface, and in evaluating his laboratory data he assumed the pesticides would be distributed in the upper 2·5 cm of soil, and equated 2·7 ppm with 1 kg/ha.

The Federal Biological Research Centre for Agriculture in Germany adopted guideline No. IV-4–1 in 1986, which recommends assuming that a pesticide will infiltrate the soil to a depth of 5 cm, and that 1·33 ppm in an earthworm toxicity test would be equivalent to 1 kg/ha in soil (density assumed to be 1·5 g/cm^3).

SOIL LAYERS SAMPLED IN STUDIES ON EARTHWORMS

Several researchers have compared concentrations of pesticides or PCBs in earthworms to those in soil, and we may infer that these researchers considered the sampling depth they selected appropriate for estimating the exposure of earthworms. Sampling depth may be critical to their conclusions; if a pesticide were limited to the uppermost 2 cm of soil, for example, then sampling from the uppermost 2 cm would yield a concentration ten times as great as if 20 cm were sampled. Gish (1970) sampled soil to a depth of 1 in (2·5 cm) in a study of organochlorine insecticides in earthworms and other soil invertebrates on farms. In a study of PCBs in earthworms, Kreis *et al.* (1987) sampled the 5–10 cm layer, as representative of the plow layer. Davis (1968) sampled soil to a depth of 12 cm in a survey of organochlorine insecticides in earthworms and macrofauna on 12 English farms. Fifteen cm (6 in) of soil was sampled in a study of the persistence of aldrin in soil and earthworms (Korschgen, 1971) and in a survey of organochlorine insecticides in earthworms and soil in India (Yadav *et al.*, 1981). Wheatley and Hardman (1968) sampled the uppermost 22 cm of soil in a study of organochlorine insecticides in earthworms in arable soils.

In studies of plowed fields the choice of 15 cm seems appropriate. That is the layer within which soil is frequently mixed during plowing, it is the soil layer containing most of the organic matter and, hence, food for earthworms, and it is where most earthworms will be found when actively feeding. Korschgen (1971) suggested that if a plow layer weighs about 2 million lbs. per acre (~2·2 million kg/ha) then 1 ppm in a plow layer is equivalent to 2 lbs./a (2·2 kg/ha). Van Leemput (1989) equated 1 ppm to 2 kg/ha. Lichtenstein *et al.* (1960) used the weights of the particular soil types they were studying to conclude that 1 lb. of an insecticide applied per acre (~2·2 kg/ha) to a depth of 6 in (15 cm) would yield 0·53–0·61 ppm in loamy soils and 1·5–1·8 ppm in muck soils. Brady (1974) states that an acre-foot of mineral soil weighs from 3 to 4 million lbs. (dry), which gives a conversion factor of 1 ppm being equivalent to 1·7–2·2 kg/ha of a 15-cm layer.

DISTRIBUTION OF PESTICIDES IN SOIL

The distribution of pesticides in soil is dependent on the method of application, the chemical nature and formulation of the pesticide, the time since application, and many site-specific variables. Relevant site-specific variables include soil texture, organic matter content, moisture and pH, and the weather (Nash, 1988). After application the pesticide may be in a film on the soil surface or, at the other extreme, if the pesticide is relatively soluble and the soil wet, it may be dispersed into the profile. If a pesticide is applied as a band in the soil and covered, then the variation perpendicular to the bands may be as great as the variation with depth. The concentration in the band itself would be extremely high, at least until the pesticide dispersed.

Most of the published studies on pesticide distribution in soil have examined relatively immobile and persistent pesticides. DDT, when applied to the soil surface and measured in the soil 17 months later, remained mainly in the uppermost 7·5 cm soil layer (Lichtenstein, 1958). When dieldrin was incorporated into the uppermost 7·5 cm of soil, no movement within the profile was detected after two and half years (Freeman *et al.*, 1975). Dieldrin, lindane, and DDT were found to be relatively immobile in soil throughout a fifteen-year study (Voerman and Besemer, 1970). Herbicides tend to be more mobile, but when evapotransporation exceeds rainfall, as is common in temperate regions after herbicide application, the maximum concentrations will generally be only a few centimeters deep (Hance, 1988). Studies of the movement of parathion in soil have had different results, depending on soil conditions; sometimes parathion remained primarily within the upper 5 cm, and sometimes much of it leached to a depth of about 15 cm (Mulla *et al.*, 1981). When an emulsion of either chlorpyrifos or diazinon was applied to turf, about half was deposited on the grass and half in soil (Kuhr and Tashiro, 1978). A greater proportion of these insecticides reached the soil when applied as granules. Even after 44 days, almost all of these insecticides remained in the uppermost 1·3 cm of soil.

If we consider relatively insoluble pesticides a few months after application to unplowed soil, we conclude that the pesticides will generally be concentrated in a soil layer that is from one to several cm thick. When interpreting results from earthworm toxicity tests, however, it is probably preferable to consider the time immediately after application, when the pesticide would be most concentrated and the hazard to earthworms would presumably be the greatest. In this extreme situation, the soil layer of concern would generally be measured in millimeters rather than centimeters.

RANGE OF EARTHWORM ACTIVITY

The distribution of earthworms within a soil profile is as important to consider as is the distribution of pesticides. Various kinds of information on earthworms are relevant. The distances moved by earthworms per day, for example, have been measured under laboratory conditions. *Aporrectodea caliginosa* burrowed an average of from 2·6 to 6·2 cm per day, depending on the temperature and moisture conditions (Scheu, 1987). *Allolobophora rosea* formed an average of about 3 cm of new burrow per day (Bolton and Phillipson, 1976). *Allolobophora caliginosa*, *Allolobophora trapezoides*, and *Lumbricus rubellus* were found to burrow from 14 to 60 cm per day per g of earthworm (Martin, 1982), which is equivalent to about 12 to 27 cm per day per earthworm. Earthworm size is also relevant. The preserved lengths of adults of the trapezoides group of earthworms (*Aporrectodea trapezoides*, *tuberculata*, *turgida*), which is especially important in many agricultural soils, are generally in the range of 6 to 15 cm (Reynolds, 1977). These observations suggest that a scale of a few millimeters in soil is too small to be appropriate for active earthworms. Even if earthworm burrows are mainly horizontal, the soil layer active earthworms inhabit is at least a few centimeters thick.

Gerard (1967) studied the depth of earthworms in different seasons. Generally, most earthworms of a species were likely to be concentrated in only one of several 7·5-cm soil layers he sampled. When conditions were favorable to earthworms they were close to the surface. In March, for example, earthworms of both *Allolobophora chlorotica* and *A. caliginosa* were limited to the uppermost 5 cm. Unfavorable conditions drove the earthworms deeper, and over a period of 12 cold days in January, earthworms of *A. chlorotica* were found to retreat from the uppermost 7·5 cm to below 15 cm. However, during the time that earthworms are actively feeding near the surface and at greatest risk from exposure to pesticides, it seems reasonable to assume that the earthworms of at least some species that are relevant to the artificial soil test are within a few cm of the soil surface. This generalization is not accurate for all species of earthworms. Earthworms such as *Lumbricus terrestris* usually feed on the soil surface and could be exposed through their skin to very high concentrations of a surface-applied pesticide, but be well below the pesticide when in their burrows. However, because the artificial soil tests are designed for earthworms that primarily ingest soil, and because we should not expect a single laboratory test to be appropriate for all species of earthworms, it makes sense to consider the species of earthworms that remain in the soil.

Conclusion

It has been concluded that the distribution of pesticides in soils is highly variable, and that in the extreme case soon after application, the pesticide may be concentrated in a soil layer as thin as a few millimeters. A consideration of earthworm size and activity, however, suggests that a typical active earthworm would not confine its activities to such a narrow layer for more than a short time. The artificial soil test lasts 14 days, and the selected soil depth should be relevant to that length of time. Even if a pesticide were concentrated in a layer only a couple of millimeters thick, it would be misleading to equate the concentration in that layer with the concentration used in the artificial soil test, because the earthworms would move around and be exposed to soil having low pesticide concentrations. This means that it is not necessary to understand precisely how pesticides are distributed in soil soon after application, but rather it is necessary to select a layer of soil thought to include most of an earthworm's activity and assume that the average concentration in this layer is an appropriate measure of exposure. This layer is less than the 15 cm of plow layer that has been used in calculations in several published studies of pesticides in earthworms, but greater than the layer in which recently applied pesticides might be concentrated. For the purpose of applying laboratory results to field conditions, I suggest that we accept the approximation that a species such as *A. caliginosa* would inhabit the uppermost 5 cm of soil during 14 days of favorable weather. This is supported by Gerard's (1967) study. Based on Brady's (1974) estimate of an acre-foot (0·4 ha by 30 cm deep) of soil weighing 3 to 4 million lbs. (1·4 to 1·8 million kg), this 5-cm layer would weigh about 0·6 million kg/ha. An application of 1 kg/ha of a pesticide would then be equivalent to 1·67 ppm (dry wt) in the artificial soil test.

References

BOLTON, P.J., PHILLIPSON, J. (1976) Burrowing, feeding, egestion and energy budgets of *Allolobophora rosea* (Savigny) (Lumbricidae). *Oecologia (Berlin)* **23**, 225–245.

BRADY, N.C. (1974) *The Nature and Properties of Soils*, New York: Macmillan Publishing Co., Inc. p. 53.

DAVIS, B.N.K. (1968) The soil macrofauna and organochlorine insecticide residues at twelve agricultural sites near Huntingdon. *Annals of Applied Biology* **61**, 29–45.

EUROPEAN ECONOMIC COMMUNITY. (1985) EEC Directive 79/831, Annex V, Part C. Methods for the determination of ecotoxicity. Level I.C(II) 4: Toxicity for earthworms. Artificial soil test. DG XI/128/82.

FREEMAN, H.P., TAYLOR, A.W., EDWARDS, W.M. (1975) Heptachlor and dieldrin disappearance from a field soil measured by annual residue determinations. *Journal of Agriculture and Food Chemistry* **23**, 1101–1105.

GERARD, B.M. (1967) Factors affecting earthworms in pastures. *Journal of Animal Ecology* **36**, 235–252.

GISH, C.D. (1970) Organochlorine insecticide residues in soils and soil invertebrates from agricultural lands. *Pesticides Monitoring Journal* **3**, 241–252.

HANCE, R.J. (1988) Adsorption and bioavailability. In: *Environmental Chemistry of Herbicides*, Vol 1, R. Grover (ed), Boca Raton, Fl: CRC Press, Inc., 1–19.

HARRASS, M.C., EIRKSON, C.E., III,; NOWELL, L.H. (1991) Role of plant bioassays in FDA Review: Scenarios for terrestrial exposures. In: *Plants for Toxicity Assessment*, Vol 2, J.W. Gorsuch, W.R. Lower and K.R. St. John (eds), ASTM STP 115, American Society for Testing and Materials, Philadelphia, Pa. (in press).

HEIMBACH, F. (1992) Correlation between data from laboratory and field tests for investigating the toxicity of pesticides to earthworms. *4th ISEE*, Avignon. (in press).

HOERGER, F., KENAGA, E. E., (1972) Pesticide residues on plants: correlation of representative data as a basis for estimation of their magnitude in the environment. In: *Environmental Quality and Safety*, Vol. 1, New York: Academic Press, Inc., pp. 9–27.

KORSCHGEN, L.J. (1971) Disappearance and persistence of aldrin after five annual applications. *Journal of Wildlife Management* **35**, 494–500.

KREIS, B., EDWARDS, P., CUENDET, G., TARRADELLAS, J. (1987) The dynamics of PCBs between earthworm populations and agricultural soils. *Pedobiologia* **30**, 379–388.

KUHR, R.J., TASHIRO, H. (1978) Distribution and persistence of chlorpyrifos and diazinon applied to turf. *Bulletin of Environmental Contamination and Toxicolog* **20**, 652–656.

LICHTENSTEIN, E.P., DE PEW, L.J., ESHBAUGH, E.L., SLEESMAN, J.P. (1960) Persistence of DDT, aldrin, and lindane in some midwestern soils. *Journal of Economic Entomology* **53**, 136–142.

LICHTENSTEIN, E.P. (1958) Movement of insecticides in soils under leaching and nonleaching conditions. *Journal of Economic Entomology* **51**, 380–383.

MARTIN, N.A. (1982) The interaction between organic matter in soil and the burrowing activity of three species of earthworms (Oligochaeta: Lumbricidae). *Pedobiologia* **24**, 185–190.

MULLA, M.S., MIAN, L.S., KAWECKI, J.A. (1981) Distribution, transport, and fate of the insecticides malathion and parathion in the environment. *Residue Reviews* **81**, 1–172.

NASH, R.G. (1988) Dissipation from soil. In: *Environmental Chemistry of Herbicides*, Vol 1, R. Grover (ed), Boca Raton, Fl: CRC Press, Inc., 131–169.

ORGANIZATION FOR ECONOMIC COOPERATION AND DEVELOPMENT. (1984) Guideline for testing chemicals no. 207. Earthworm, acute toxicity tests. Adopted 4 April 1984.

PIŽL, V. (1988) Interactions between earthworms and herbicides. 1. Toxicity of some herbicides to earthworms in laboratory tests. *Pedobiologia* **32**, 227–232.

REYNOLDS, J.W. (1977) *The Earthworms (Lumbricidae and Sparganophilidae) of Ontario*, Toronto: Life Sciences Miscellaneous Publications, Royal Ontario Museum, 141 pp.

SCHEU, S. (1987) The role of substrate feeding earthworms (Lumbricidae) for bioturbation in a beechwood soil. *Oecologia(Berlin)* **72**, 192–196.

U.S. ENVIRONMENTAL PROTECTION AGENCY. (1986) Hazard Evaluation Division Standard Evaluation Procedure Ecological Risk Assessment. EPA 540/9–85–001.

VAN LEEMPUT, L., SWYSEN, E., WOESTENBORGHS, R., MICHIELSEN, L., MEULDERMANS, W., HEYKANTS. (1989) On the terrestrial toxicity of the fungicide Imazalil (Enilconazole) to the earthworm species *Eisenia foetida. Ecotoxicology and Environmental Safety* **18**, 313–320.

VOERMAN, S., BESEMER, A.F.H. (1970) Residues of dieldrin, lindane, DDT, and parathion in a light sandy soil after repeated application throughout a period of 15 years. Journal of Agriculture and Food Chemistry, **18**, 717–719.

WHEATLEY, G.A., HARDMAN, J.A. (1968) Organochlorine insecticide residues in earthworms from arable soils. *Journal of Agriculture and Food Chemistry* **19**, 219–225.

YADAV, D.V., MITTAL, P.K., AGARWAL H.C., PILLAI, M.K.K. (1981) Organochlorine insecticide residues in soil and earthworms in the Delhi Area, India, August-October, 1974. *Pesticides Monitoring Journal* **15**, 80–85.

Behaviour as a Source of Earthworm Susceptibility to Ecotoxicants

A.D. TOMLIN

London Research Centre, Agriculture Canada, 1400 Western Rd., London, Canada N6G 2V4

Abstract

Earthworm behaviour may not only respond to pesticide and heavy metal ecotoxicants as reduced growth and reproductive rates, but characteristic species behaviour patterns may render some worm species more exposed or susceptible to ecotoxicants. This would suggest that orthodox toxicity testing alone will not be adequate to estimate earthworm risk to ecotoxicants. There is the further concern that earthworm species that translocate soil from the depths to the surface or litter from the surface to depth (anecics) may remobilize ecotoxicants into the food chain.

Introduction

Behaviour may be defined as anything an organism does to adjust to or interact with its environment. Behaviour may be both inherited and learned. Maintenance activities such as locomotion and feeding keep an animal alive, but have little effect on conspecifics; communicatory activities are concerned with sending information to, and influencing the activities (*eg*. reproduction) of conspecifics. What data there are on interaction of earthworm behaviour with ecotoxicants is usually related to interference of maintenance activities rather than influencing subtle behaviour changes towards conspecifics that could have serious consequences for population maintenance (eg. changes in reproductive behaviour).

Although there is some evidence that earthworms may be able to learn simple behavioral routines in laboratory mazes (Howell, 1974; Rosenkoetter and Boice, 1975; McManus and Wyers, 1979), there is only a little tantalizing evidence that earthworm populations or individual earthworms change their behaviour in order to minimize their exposure to ecotoxicants occurring at sublethal concentrations. I am aware of only Morgan and Morgan's (1988) study hinting that earthworm populations develop resistance (physiological, biochemical or behavioral) to ecotoxicants, although one suspects this must happen more widely. Therefore, I am generally limiting this discussion to differences in behaviour characteristic of different earthworm species of northern temperate regions.

Differences in earthworm behaviour have been recognized since at least Darwin's (1881) classic treatise that only obliquely discusses earthworm speciation as a source of behavioral differences. Bouché's (1977) ecological classification of earthworms using the terms endogenic, epigaeic, and anecic, to describe earthworm behaviour as suites of morphological features and activities that distinguished different species, gave ecologists and agricultural scientists a ready method for categorizing earthworm behaviour. Satchell (1980) described earthworm species in terms of r and K selection, that implies obvious differences in behavioral characteristics for earthworm species at each end of the r and K spectrum. Biologists intuitively recognize that significant behavioral differences exist even amongst closely related animal species. Presumably this is an evolutionary mechanism for partitioning resources more efficiently. Consequently different species of earthworms, physically occupying the same pedon, for example, actually occupy different niches within the pedon. Implicitly, different niches suggest different responses to any ecotoxicological compounds that may be in the vicinity, because an earthworm's risk of exposure will be a function of habitat (niche), and its behaviour-activity patterns within that habitat. Collection of several different species of earthworms from a particular site or plot should remind researchers that each species probably has very different processes and frequencies of interactions with the soil and litter fabric that may greatly affect each species' exposure risk to ecotoxicants.

Behavioral differences between earthworm species are also responsible for the different distributions of ^{137}Cs (derived from atmospheric nuclear weapons testing) through the soil profile (Tomlin *et al.*, 1991). Although ^{137}Cs is usually only at very low non-hazardous concentrations, the finding does suggest that earthworms could have several mechanisms which are functions of their different species-dependent behavioral repertoires for remobilizing environmentally hazardous materials. These materials may have been assumed to be stable in waste disposal sites (soil 'sinks'), but the possibility should be considered that the integrity of these disposal sites could be breached by earthworm activity. Because earthworms are a source of food for many species of animals, there follow many opportunities for any ecotoxicant taken up or bioaccumulated by earthworms to be transported to other parts of the food chain.

Behavioral factors that modify earthworm response to ecotoxicants

GENERAL CONSIDERATIONS

The weakness of classical toxicity testing as a mechanism of certifying the safety (or hazard) of a particular compound is that it fails to address the potential ecological impact of that compound, particularly if large amounts have been disseminated over wide areas (Reinecke and Venter, 1985). Additionally, there has been little effort to incorporate behavioral vulnerability of certain species into ecotoxicological protocols.

Virtually any significant biological characteristic of any earthworm species could be considered a factor in affecting its susceptibility to ecotoxicants, but relatively few characteristics other than changes in earthworm numbers or biomass have been tested in the field for agricultural chemicals (Edwards and Thompson, 1973) or heavy metals (Carter *et al.*, 1980; Ireland and Richards, 1981). Factors such as growth and reproduction have been shown to be affected by ecotoxicants in laboratory studies (Venter and Reinecke, 1988), and Edwards and Thompson (1973) reviewed the effects of agricultural chemicals on earthworms with respect to increased mortality and decreased activities under agricultural conditions. Bouché's (1977) ecological divisions for earthworm behaviour provide a simple format with which to discuss earthworm susceptibility to ecotoxicants. Satchell (1980) tabulated the characteristics of Bouché's three divisions, some of which are summarized in Table 1. Several of these traits including size, maturation and respiration rates could significantly influence both maintenance and communicatory behaviour that could change earthworm response to ecotoxicants.

Table 1. Summary of some of the characteristics that differentiate Bouché's three ecological divisions of lumbricids.

Trait	Epigeics	Endogeics	Anecics
Burrowing muscles	Reduced	Developed	Strongly developed
Weight	10–30 mg	10–1100 mg	200–1100mg
Fecundity	High	Limited	Limited
Maturation	Rapid	Moderate	Moderate
Respiration rate	High	Feeble	Modest
Food size	Mesophage	Microphage	Macrophage

THE PARTICULAR VULNERABILITY OF *L. TERRESTRIS*

The surface mating and feeding of *Lumbricus terrestris* L. makes it especially susceptible to chemicals applied to vegetation and granular pesticide formulations applied to the soil surface (Lofs-Holmin, 1980; Ebing *et al.*, 1985); its vertical burrowing activity and surface casting activity mean that it is capable of translocating pollutants from the soil depths to the surface and vice versa. The wide geographical occurrence and frequently high abundance of this species in temperate agricultural regions should make it a special target for concern and study. Should this species suffer extensive mortality from ecotoxicants, its relatively slow growth and reproductive rate would reduce its recolonization rates.

There are significant economic consequences in Ontario, Canada, where there has been a long-standing and profitable bait-worm industry centred on the harvesting of *L. terrestris* (Tomlin, 1983). If worms have been exposed to pollutants, subsequent worm harvesting by hand-picking followed by high density storage and shipping may induce further stress, thus increasing shelf mortality and incurring considerable expense for both worm harvesters and distributors.

Case studies with agricultural chemicals

Under leadership of OECD scientists, standardized techniques using *Eisenia fetida* (Savigny) and an artificial soil were developed (Heimbach and Edwards, 1983) to reduce registration costs for pesticide manufacturers and expedite the registration process. Although this test is a useful benchmark, it should be recognized that there are inherent differences amongst earthworm species in response to toxicants as measured by dosage-response testing techniques (Davey, 1963). Under field conditions, the problem is further compounded by the different habitats and behaviour for different species of earthworms that may change their frequency, intensity and length of exposure to the ecotoxicant.

Soil type (including proportions of clay, mineral, and humus) affects the expression of toxicity of (agricultural) organic chemicals (Harris, 1966, Edwards, 1974). However, distribution of earthworm species is also affected by soil type (Guild, 1948 and 1951), and because earthworm behaviour is species dependent, the interaction of effects may complicate the interpretation or prediction of toxicity of an ecotoxicant, particularly under field conditions.

INSECTICIDES

One of the earliest demonstrations of insecticide effects on earthworm behaviour was Baker's observation (1946) that leaves were accumulating under elm trees sprayed with DDT because earthworm populations had been severely reduced and were no longer removing leaves from the surface. By 1962, Carson's book, *Silent Spring*, although sensationalized in the media, publicized the fact that DDT was entering the food chain and killing birds. The evidence rested on the suggestion that earthworms were accumulating DDT residues and passing the residues up the food chain. For the last 2 decades it has been common practice for pesticide registration officials to request that earthworm toxicity studies be undertaken before allowing sale of new pesticides (Goats and Edwards, 1988).

Unfortunately we have suffered from the fact that most studies on earthworm burdens of insecticide residues undertaken in the 1960's and 1970's did not differentiate amongst the several earthworm species that may have been collected for analysis (Ebing *et al.*, 1985). These authors recognized that an earthworm's ability to pick up ecotoxicants was species (hence habit) dependent, but when they inspected data of many early workers, they were usually unable to relate species behaviour with different body burdens of insecticide residues. Workers often quoted residue values for a particular insecticide for *L. terrestris* for example, but listed only one value for all the other species that may have been collected along with *L. terrestris*. One presumes that methodological problems due to the fact that *L. terrestris*, both large and abundant at many of the collection sites, provided plenty of tissue for residue analysis, whereas the other collected species were small and rarer, making residue analyses of these species more difficult.

Two tropical species of earthworms, *Hyperiodrilus* spp. and *Eudrilus* spp. responded to differences in preference in vegetation cover, with *Eudrilus* dominating cultivated plots and *Hyperiodrilus* declining after forest clearing (Madge, 1969). There was evidence that DDT treatments on cultivated soils, although reducing surface casting activity significantly, did not change worm numbers in the soil, suggesting a repellant effect or avoidance reaction by the worms (Cook *et al.*, 1980).

Curl *et al.* (1987) demonstrated in laboratory experiments that the level of bioaccumulation of cypermethrin was nearly 4 times higher in *A. caliginosa* (30-fold accumulation over soil levels) than in *L. terrestris* (8-fold accumulation). The authors interpreted this as a result of the geophagous habit of *A. caliginosa* that fully exposed it to the insecticide soil treatment; *L. terrestris*, by contrast, lived in permanent burrows and fed on litter collected at the soil surface, thus reducing its exposure in this case to insecticide treatment.

HERBICIDES

Lofs-Holmin (1980) noted in a laboratory study that trichloroacetic acid (TCA) was toxic to *Allolobophora rosea* (Savigny), *A. chlorotica* (Savigny), and *L. terrestris*, and caused retarded maturity in *A. caliginosa* (Savigny) at rates of 20 and 40 kg/ha, compared to the 10–30 kg/ha for field applications in Sweden.

Tomlin (unpublished data, 1990) found that after 8 years of intensive row-cropping (maize and soya beans), herbicide treatments (metolachlor and atrazine for maize, metolachlor and metribuzin for soy beans) had significantly more juvenile *Aporrectodea turgida* than machine cultivated plots; adult *A. turgida* and adult and juvenile *L. terrestris* numbers were not significantly different for the same treatments. I interpret these results to mean that different species, and even different life stages of the same species (juvenile *L. terrestris* are endogeic rather than anecic) have differing susceptibilities to these agricultural practices (*ie.* machine cultivation is more damaging to some species and life stages of earthworms than recommended herbicide treatment rates).

Different formulations of the same active ingredients can also change the susceptibility of earthworms to ecotoxicants. Again, the surface-active habit of *L. terrestris* makes it particularly vulnerable. Tomlin (unpublished data, 1975), was informed by a bait wholesaler in the Ontario region that *L. terrestris* worms were dying a few days after being picked from two particular golf courses; a third golf course harvested in the same area was problem-free (the ability to segregate worm harvests from different courses was possible because each worm picker had a code number that went with each collection of 1000 worms to point of retail sale). The only obvious difference between golf courses was that the 'problem' courses had recently switched from water-formulated, sprayed herbicide (1·46 kg/ha of 2,4-D-amine + 0·07 kg/ha of dicamba) and separate granular fertilizer (145 kg/ha of 32-5-3 of N-P-K) treatment to a product that formulated the herbicide at the same rates onto the fertilizer granules allowing combined application of herbicide and fertilizer treatment. In a laboratory test with *L. terrestris* that compared an untreated control, fertilizer alone, and the combined herbicide-fertilizer formulation, 84% of the worms in the combined formulation treatment were dead after 21 days of exposure; there was no mortality in the other two treatments. The manufacturer of the herbicide fertilizer formulation had previously tested 'small red garden worms' (species unrecorded) at recommended treatment rates, but found no mortality after 21 days. Although this work was never followed up, I would assume that the source of the intoxication is surface activity of *L. terrestris* that exposes it to surface-applied granules of herbicide.

FUNGICIDES

Benomyl toxicity was originally detected as an earthworm behavioral change when mulched grass was noted accumulating in regularly mown orchards treated with benomyl (Kennel, 1972) in a situation similar to that observed above with DDT and elm trees. Lofs-Holmin (1980) noted in a laboratory study that there were different growth responses for different earthworm species to benomyl; species responses also varied by amount of sand or clay in the soil mixture. It would appear that clitellum development (thus sexual maturity and reproductive behaviour) could be delayed by exposure to sublethal doses of benomyl. Haque and Ebing (1983) clearly stated in their discussion of Lofs-Holmin's data, that 'where mixed populations of earthworms are expected to be present the effect of pesticides may differ within species depending on the living and feeding habits of worms'.

Keogh and Whitehead (1975) found no difference in earthworm numbers (no species given) between fungicide-treated plots and unsprayed plots, but significantly reduced casting activity and higher litter residues in the treated plots compared to unsprayed plots. The authors concluded that benomyl and carbendazim may be materials that have repellant action or inhibit feeding of earthworms.

Near a long-closed sawmill in Finland, chlorinated phenols used for wood preservation were found to concentrate more in geophagous *Aporrectodea caliginosa* than in *Lumbricus* spp. because it was presumably exposed to lower concentrations of the preservative at the soil surface (Knuutinen *et al.*, 1990).

Heavy metals

Although earthworms are an obvious candidate for measuring non-target effects of pesticides in agricultural soils, earthworms have also been used to measure heavy metal contamination of the environment from industrial activities (Ireland, 1983) and leaded fuels from motor vehicle operation (Gish and Christiansen, 1973).

Tissue studies of earthworms revealed that the posterior alimentary canal contained the highest concentrations and proportions of heavy metal (Pb and Cd) burdens (Ireland, 1975; Morgan and Morgan, 1990). Morgan and Morgan (1988) had some evidence that lead sequestration was affected by calcium-rich chloragosome granules, and that there was a tolerance mechanism to lead that had a possible genetic basis (*ie.* possible resistance). Because these physiological response mechanisms to heavy metals are demonstrably species specific, and the route by which heavy metals may reach the soil vary (*eg.* fly ash deposition, sludge disposal on soils, mine and smelter wastes), different earthworm behaviour patterns will result in different species vulnerabilities to this group of ecotoxicants.

Discussion

It is important that ecotoxicants be subjected to testing that more completely reveals their ecological hazard or impact. As Callahan (1988) emphasized, testing of sublethal effects on growth, reproduction and neurophysiological responses will be required to connect laboratory and field effects of ecotoxicants. The contact and soil tests, by themselves, are insufficient for a full interpretation of ecological impact. I would add that (where conditions warrant), the tests should be expanded to include other species besides *E. fetida*. *E. fetida* seems more tolerant of ecotoxicants (Dean-Ross, 1983), and its habitat and behaviour are insufficiently representative of many other more widely distributed and economically important earthworm species for *E. fetida* to be an appropriate representative of all earthworm species for studying the ecological impact of ecotoxicants. However, its role as proxy for toxicity testing now seems assured (Heimbach, 1985). Provided the weaknesses of the proposed test protocols are understood by the earthworm research community and regulatory officials, there could be improvements in research efficiency and experimental design advantages by standardizing on one test species. Research or debate of this issue should continue and not be stifled by the standardization of protocols.

Lofs-Holmin and Bostrom's (1988) literature survey of pesticide testing for earthworms demonstrates that we need to be more critical of our own and other published work to ensure that sufficient information on experimental conditions for soil and weather parameters is included, as well as biological data, to improve our ability to make rational decisions concerning the ecological consequences of using a toxicant. Standardizing on one test species could well improve the quality of research and reporting of precisely these issues. It should not be too difficult to determine whether earthworms have developed resistance to ecotoxicants, and the introduction of rapid, standardized techniques for earthworm toxicity testing should simplify this type of research.

There is further need to study the behaviour of earthworm species and their interactions with ecotoxicants, because soil acts as both an inadvertent sink (*eg.* pesticide run-off from agricultural activities, lead residues derived from anti-knock compounds near highways) and as a designed sink for toxic substances (waste disposal sites). Toxic substances from these sites could be picked up by earthworms and subsequently returned to the environment via the food chain. Species specific behaviour patterns and abundance should be critical considerations in determining whether a particular species of earthworm would be capable of remobilizing toxic substances from soil sinks. It is apparent that until we know more about the specific interactions of earthworms with soil microfabric and soil aggregates, it will be very difficult to set experimental testing protocols with confidence for field studies.

Acknowledgements

The author acknowledges the assistance of J. Miller and J. Mitton in preparation of the manuscript.

References

BAKER, W.L. (1946) DDT and earthworm populations. *Journal of Economic Entomology* **38,** 404–405.

BOUCHÉ, M.B. (1977) Strategies lombriciennes. In: Soil organisms as components of ecosystems, U. Lohm and T. Persson (Eds.). *Ecological Bulletins* (Stockholm) **25**, 122–132.

CALLAHAN, C.A. (1988) Earthworms as ecological assessment tools. In: *Earthworms in waste and environmental management*, C.A. Edwards and E.F. Neuhauser (Eds.), SBP Academic Publishing, The Hague, pp. 295–301.

CARSON, R. (1962) *Silent Spring*. Houghton Mifflin, New York. 304 pp.

CARTER, A., KENNEY, E.A., GUTHRIE, T.F. (1980) Earthworms as biological monitors of changes in heavy metals in an agricultural soil in British Columbia. Soil Biology as related to land use practices. *Proceedings of the VII International Soil Zoology Colloquium, Syracuse, N.Y.* D.L. Dindal (Ed.), United States Environmental Protection Agency, Washington, D.C., Office of Pesticides and Toxic Substances. EPA-560/13–80–038. pp. 344–357.

COOK, A.G., CRITCHLEY, B.R., CRITCHLEY, U., PERFECT, T.J., YEADON, R. (1980) Effects of cultivation and DDT on earthworm activity in a forest soil in the sub-humid tropics. *Journal of Applied Ecology* **17**, 21–29.

CURL, E.A., EDWARDS, P.J., ELLIOTT, C., LEAHEY, J.P. (1987) The conjugation and accumulation of metabolites of cypermethrin by earthworms. *Pesticide Science* **20**, 207–222.

DARWIN, C. (1881) *The formation of vegetable mould through the action of worms with observations on their habits*. Murray, London. 298 pp.

DAVEY, S.P. (1963) Effects of chemicals on earthworms: a review of the literature. *U.S. Fish & Wildlife Service Sp. Sci. Rep. Wildlife #74* **2**, 1–20.

DEAN-ROSS, D. (1983) Methods for the assessment of the toxicity of environmental chemicals to earthworms. *Regulatory Toxicology and Pharmacology* **3**, 48–59.

EBING, W., PFLUGMACHER, J., HAQUE, A. (1985) Survey of the suitability of certain plants and animals for use as index organisms in monitoring soil pollution by organic chemicals [with special reference to earthworms]. *Plant Research and Development* **22**, 37–84.

EDWARDS, C.A. (1974) Soil types and agricultural pest control. *Soil Survey Technical Monograph* **4**, 83–90.

EDWARDS, C.A., THOMPSON, A.R. (1973) Pesticides and the soil fauna. *Residues Reviews* **45**, 1–81.

GISH, C.D., CHRISTENSEN, R.E. (1973) Cadmium, nickel, lead, and zinc in earthworms from roadside soil. *Environmental Science and Technology* **7**, 1060–1062.

GOATS, G.C., EDWARDS, C.A. (1988) The prediction of field toxicity of chemicals to earthworms by laboratory methods. In: *Earthworms in waste and environmental management*, C.A. Edwards and E.F. Neuhauser (Eds.), SBP Academic Publishing, The Hague, pp. 283–294.

GUILD, W.J.MCL (1948) Effect of soil type on populations. *Annals of Applied Biology* **35**, 181–192.

GUILD, W.J.MCL (1951) The distribution and population density of earthworms (Lumbricidae) in Scottish pasture fields. *Journal of Animal Ecology* **20**, 88–97.

HAQUE, A., EBING, W. (1983) Toxicity determination of pesticides to earthworms in the soil substrate. *Journal of Plant Diseases and Protection* **90**, 395–408.

HARRIS, C.R. (1966) Influence of soil type on the activity of insecticides in soil. *Journal of Economic Entomology* **59**, 1221–1225.

HEIMBACH, F., EDWARDS, P.J. (1983) The toxicity of 2-chloroacetamide and benomyl to earthworms under various test conditions in an artificial soil test. *Pesticide Science* **14**, 634–637.

HEIMBACH, F. (1985) Comparison of laboratory methods, using *Eisenia foetida* and *Lumbricus terrestris*, for the assessment of the hazard of chemicals to earthworms. *Zeitschrift fur Pflanzenkrankheiten und Pflanzenschutz* **92**, 186–193.

HOWELL, D.N. (1974) The worm turns: an investigation of experimentation on the learning abilities of earthworms. *Megadrilogica* **1**, 1–6.

IRELAND, M.P. (1975) Metal content of *Dendrobaena rubida* (Oligochaeta) in a base metal mining area. *Oikos* **26**, 74–79.

IRELAND, M.P. (1983) Heavy metal uptake and tissue distribution in earthworms. In: *Earthworm Ecology*, J.E. Satchell (Ed.). Chapman and Hall, London. pp. 247–265.

IRELAND, M.P., RICHARDS, K.S. (1981) Metal content, after exposure to cadmium, of two species of earthworms of known differing calcium metabolic activity. *Environmental Pollution (A)* **26**, 69–78.

KENNEL, W. VON. (1972) Schadpilze als Objekte integrierter Pflanzenschultzmabnahmen im Obstbau. *Zeitschrift fur Pflanzenkrankheiten und Pflanzenschutz* **79**, 400–406.

KEOGH, R.G., WHITEHEAD, P.H. (1975) Observations on some effects of pasture spraying with benomyl and carbendazim on earthworm activity and litter removal from pasture. *New Zealand Journal of Experimental Agriculture* **3**, 103–104.

KNUUTINEN, J., PALM, H., HAKALA, H., HAIMI, J., HUHTA, V., SALMINEN, J. (1990) Polychlorinated phenols and their metabolites in soil and earthworms of sawmill environment. *Chemosphere* **20**, 609–623.

LOFS-HOLMIN, A. (1980) Measuring growth of earthworms as a method of testing sublethal toxicity of pesticides. *Swedish Journal of Agricultural Research* **10**, 25–33.

LOFS-HOLMIN, A., BOSTROM, U. (1988) The use of earthworms and other soil animals in pesticide testing. In: *Earthworms in waste and environmental management*, C.A. Edwards and E.F. Neuhauser (Eds.), SBP Academic Publishing, The Hague, pp. 303–313.

MADGE, D.S. (1969) Field and laboratory studies on the activities of two species of tropical earthworms. *Pedobiologia* **9**, 188–214.

MCMANUS, F. E., WYERS, E.J. (1979) Olfaction and selective association in the earthworm, *Lumbricus terrestris*. *Behavioral and Neural Biology* **25**, 39–57.

MORGAN, J.E., MORGAN, A.J. (1988) Calcium-lead interactions involving earthworms. Part 2: the effect of accumulated lead on endogenous calcium in *Lumbricus rubellus*. *Environmental Pollution* **55**, 41–54.

MORGAN, J.E., MORGAN, A.J. (1990) The distribution of cadmium, copper, lead, zinc and calcium in the tissues of the earthworm *Lumbricus rubellus* sampled from one uncontaminated and four polluted soils. *Oecologia* **84**, 559–566.

REINECKE, A.J., VENTER, J.M. (1985) Influence of dieldrin on the reproduction of the earthworm *Eisenia fetida* (Oligochaeta). *Biology and Fertility of Soils* **1**, 39–44.

ROSENKOETTER, J.S., BOICE, R. (1975) Earthworm pheromones and T-maze performance. *Journal of Comparative and Physiological Psychology* **88**, 904–910.

SATCHELL, J.E. (1980) r Worms and K worms: a basis for classifying lumbricoid earthworm strategies. In: *Soil Biology as related to land use practices. Proceedings of the VII International Soil Zoology Colloquium, Syracuse, N.Y.* D.L. Dindal (Ed.), United States Environmental Protection Agency, Washington, D.C., Office of Pesticides and Toxic Substances. EPA-560/13–80–038. pp. 82–93.

TOMLIN, A.D. (1983) The earthworm bait market in North America. In: *Earthworm Ecology*, J.E. Satchell (Ed.). Chapman and Hall, London. pp. 331–338.

TOMLIN, A.D., MCCABE, D., PROTZ, R. (1992) Species composition and seasonal variation of earthworms, and their effect on soil properties in southern Ontario, Canada. Soil Biol. Biochem. (In press).

VENTER, J.M., REINECKE, A.J. (1988) Sublethal ecotoxicological effects of dieldrin on the earthworm *Eisenia foetida* (Oligochaeta). In: *Earthworms in waste and environmental management*, C.A. Edwards and E.F. Neuhauser (Eds.), SBP Academic Publishing, The Hague, pp. 337–353.

The significance of chemical residues in earthworms

Uptake and Metabolism of Xenobiotics by Earthworms

J. STENERSEN

University of Oslo, Dept. of Biology, Div. of Molecular Cell Biology, P.O. Box 1050, Blindern, N-0316 Oslo 3, Norway

Abstract

Xenobiotic biotransformation such as the metabolism of pesticides has been studied in great detail in the rat and in several fish species. However, the earthworms, which are the nontarget organisms most likely to be harmed by pesticides and other xenobiotics have been offered only very few studies. Important biotransformation enzymes, such as esterases, cytochrome-P450s, glutathione transferases and some other conjugation enzymes are known to be present in the earthworm. The studies done include the toxicokinetics of a few phenols, some organophosphorus and carbamate insecticides and a pyrethroid insecticide as well as dioxin. Some of these studies will be briefly commented upon. Studies of xenobiotic metabolism is complicated by adsorption in the soil which also strongly influence the uptake and toxicity of chemicals.

Introduction

Very few studies have been carried out on biotransformation of xenobiotics in earthworms as compared to vertebrates, in spite of the fact that earthworms are a group of nontarget organisms most likely to be harmed by pollutants reaching the soil medium. Surprisingly few reports have appeared on the subject since the review published in 1984 (Stenersen 1984).

Uptake and Elimination

A compartmental model (Fig. 1), is useful to clarify the system of earthworms and soil. The polluting substance dissolved in the soil water will diffuse into the earthworm tissue, most probably by first order reaction kinetics (k_1). First order uptake kinetics from water solution have been observed for **carbofuran** in *Lumbricus terrestris* (Stenersen *et al.* 1973) and **carbofuran** and **oxamyl** in

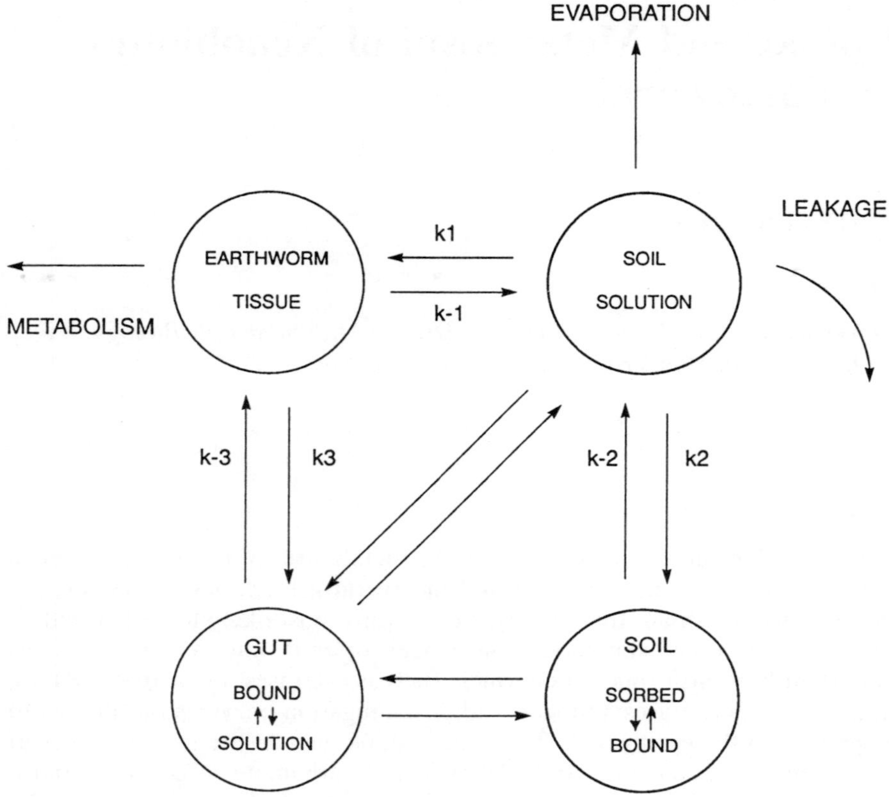

Figure 1. A compartment model describing the distribution and elimination routes of a pesticide in the system consisting of soil and earthworms.

Eisenia andrei (Stenersen and Øien 1980). The substance may also be taken up by the gut (k_3). Uptake competes with other reactions removing the substance from the soil water compartment, eg. leakage, adsorption (k_2), binding, evaporation or degradation. The worm may metabolise, secrete or excrete (k_{-3}) the substance. Secretion and excretion also take place by first order reaction kinetics as shown for **carbofuran** and **parathion** (Stenersen 1979b). Rates of uptake strongly influences the toxicity of **chlorophenols** as shown by Gestel and Ma (1988). They found a clear relation between toxicity and octanol-water partition coefficients (Fig. 2) and concluded that the toxicity towards earthworms of these substances is dependent on the soil water concentration as well as the K_{ow}. The toxicity of the chlorophenols was independent of soil type when related to soil water concentration (rather than to mg/kg dry soil matter as is often done). The ability of different toxicants to be adsorbed greatly influences the test results in standard toxicity tests using soil or artificial soil substitutes. The results obtained by Heimbach (1984) for 19 toxicants tested on **E. foetida** in artificial soil and on moistened filter paper, may probably be explained by the different adsorption rates

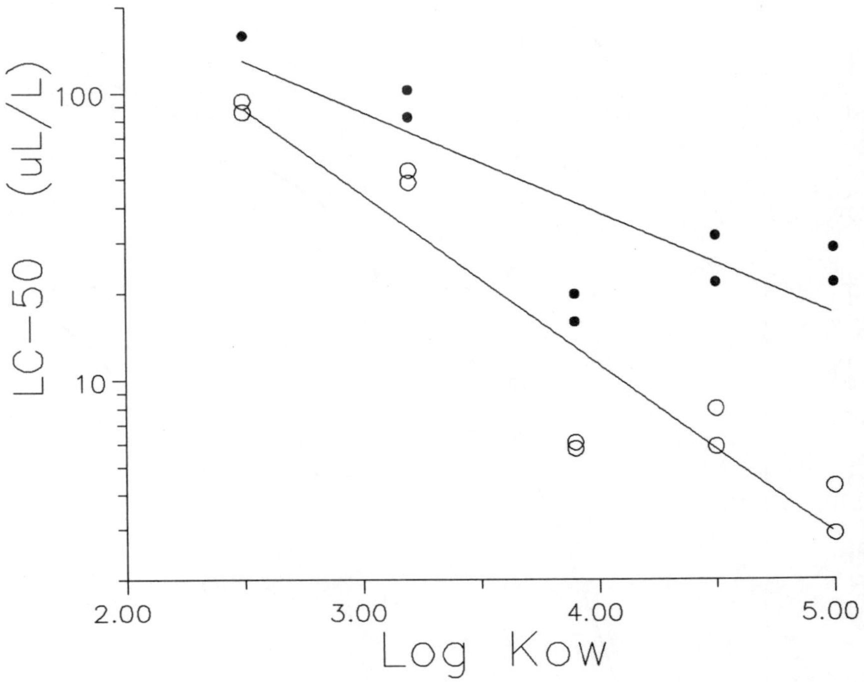

Figure 2. Toxicity (LC_{50} in µmol per L of soil solution) of chlorophenols to *E. andrei* (open symbols) and *L. rubellus* (closed symbols) as a function of the octanol- water partition coefficients. (Data from Gestel and Ma 1988) Testing was carried out at 20° (*E. andrei*) and 15°C (*L. rubellus*) in soil with 16% (w/w) water. Mortality was scored after 7 and 14 days. The LC_{50}'s were determined using 5 concentrations and 30 worms per concentration and calculated according to the 'logit' model.

between filter paper and the artificial soil. The correlation between data from these two tests was very low, the linear regression between the log LC_{50} obtained by the two methods having a correlation coefficient (r) of 0·55. The substances seem to be grouped in two categories (Fig. 3). Those close to the lower curve are likely to be strongly adsorbed.

The concentration of the chemical in the soil water will not be immediately replenished by desorption and diffusion (Lord *et al.* 1980). *L. terrestris*, which lives in permanent burrows will therefore pick up less pesticide from the soil than *Aporrectodea caliginosa*. This has been demonstrated very clearly by Curl *et al.* (1987) in the case of **cypermethrin** (Fig 4.).

An important predictor of the uptake rate and distribution between water and earthworm tissue is the octanol-water equilibrium constant (K_{ow}) of the substance. However, high K_{ow} lowers the soil water concentration by increasing the adsorption to the soil (Lord *et al.* 1980, Somasundaram *et al.* 1991) and therefore makes the substance less available to the worm. A simple linear relationship between K_{ow} and uptake to the worms from soil is therefore not to be expected.

The balance between the rate of uptake and secretion/excretion strongly

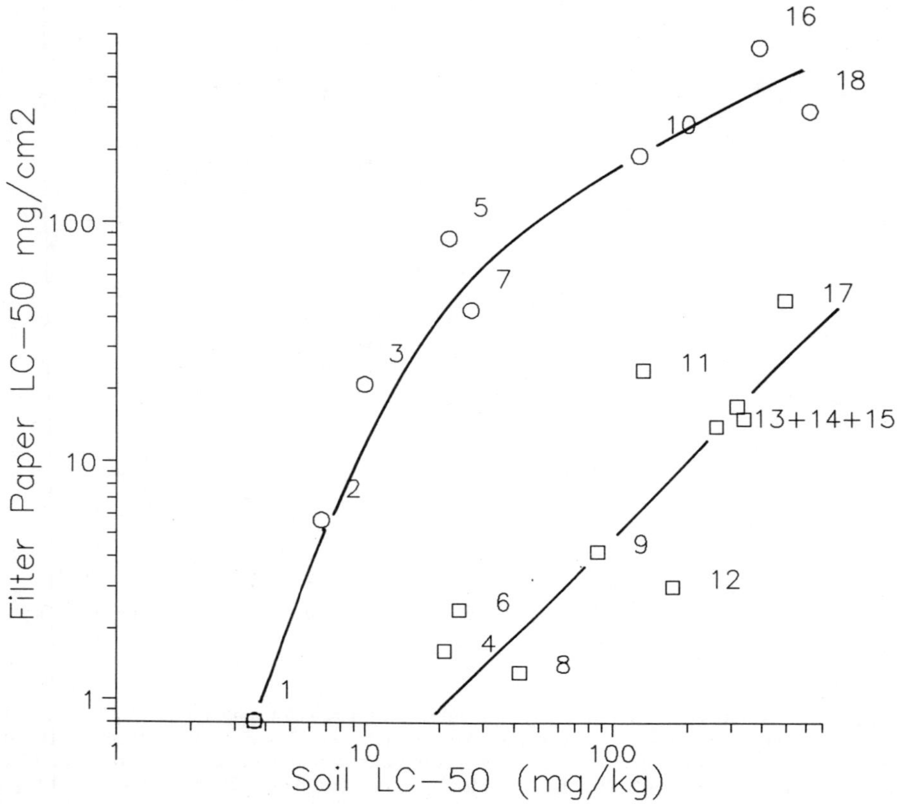

Figure 3. Toxicity of 18 substances in the Artificial Soil Test and the Filter Paper Test (Data from Heimbach 1984). The substances are distributed along two lines. 1, methidathion; 2, endosulfan; 3, propoxur; 4, DNOC; 5, benomyl; 6, chloroacetamide; 7, fentinacetat + maneb; 8, chlordane; 9, pentachlorohenol; 10, mercaptodimethur; 11, dialiphos; 12, carbaryl; 13, ethiofencarb; 14, methaphenamiphos; 15, bupirimate; 16, potassium bromate; 17, captafol; 18, captan

influences the toxicity of a compound. Thus, the uptake rate of the more toxic **carbofuran** is ten times higher than that of **oxamyl**, but the excretion rates are similar (Stenersen and Øien 1980)

PERSISTENT PESTICIDES

The very persistent organochlorine insecticides are still available to earthworms even twenty years after soil treatment (Beyer W N and Krynitsky 1989). The decrease in concentrations, with 'half-times' of about 5·5 years (**dieldrin**), 5·7 years (**DDE**), 3·5–5·7 years (**heptachlor epoxide**), are probably similar to the dissappearance rates from the soil. The earthworms may therefore be an important route of exposure to persistent pesticides for earthworm-eating predators long after the termination of pesticide usage.

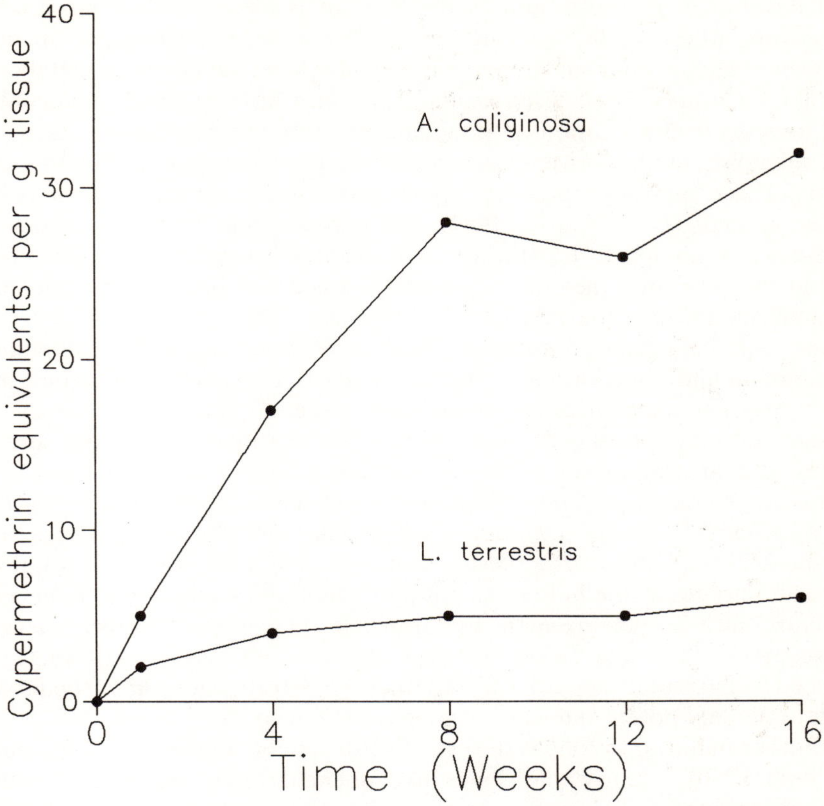

Figure 4. Content of radioactive materials originating from ^{14}C-cypermethrin in *A. caliginosa* and *L. terrestris*. (Data from Curl *et al.* 1987) The worms were kept in soil with a start concentration of 1 mg ^{14}C-cypermethrin per kg soil in 8 weeks followed by 8 weeks in pure soil. One cypermethrin equivalent corresponds to the radioactivity from 1 μg of ^{14}C-cypermethrin.

Metabolism

Earthworms may play an important role in the degradation of xenobiotics reaching the soil medium by their metabolism. But compared to mammals, fish and insects very little is known about the systems responsible for biotransformations of xenobiotics in earthworms. Complete toxicokinetic studies of pesticides have not been carried out. Some of the common biotransformation enzymes have, however, been studied, and will be described. The most important of the biotransformation routes in most animals are oxidation by the cytochrome P450 system, conjugation to glutathione or certain other conjugations, and hydrolysis.

OXIDATION

Lipophilic xenobiotics are very often subject to oxidation, and this is an important and omnipresent biotransformation reaction which is particularly well developed in insects and mammals, but is also found in most other organisms. Many of the enzymes involved are normally engaged in metabolism of endogenous substrates, but they also have the ability to metabolise foreign substrates. Specialised oxidase enzymes have evolved in mammals and insects for detoxication of secondary plant metabolites such as terpenes and alkaloids in the constant 'chemical warfare' between herbivores and plants. These enzymes represent important defence mechanisms of the plant-eating animals (Nebert *at al.* 1989), but may also catalyse detoxication of man-made xenobiotics. Earthworms are detritus eaters and have not taken part in this 'war' and they have (probably) less developed oxidative defence mechanisms towards toxicants.

The most important enzymes involved in oxidation of foreign compounds in mammals and insects are the cytochrome-P450. In mammals these enzymes are of two main categories, mitochondrial, and endoplasmatic. The enzymes connected to the mitochondrian membrane are mainly present in the adrenal cortex and are important in steroid synthesis and of little significance in xenobiotic metabolism. The endoplasmatic cyt.-P450s are currently classified as 78 individual transspecific enzymes (Gonzales 1990) based upon structure (amino acid sequences). An enzyme 'family' called 'the cyt.-PY501A family' plays an important role in the oxidation of xenobiotics with a 'flat' molecular structure such as **polyaromatic hydrocarbons** like **methylcholanthrene** and **benzoepyrene**, the insecticide **carbaryl** etc. One enzyme (cyt-P450IA1) is induced by flat molecules. **2,3,7,8–TCDD** (2,3,7,8–tetrachlorodibenzoedioxin) is an extremely potent inducer, although it is not oxidised itself.

Another important group is the cyt.- P450II family. Some of these enzymes are induced by 'clumsy' molecules like **phenobarbital, aldrin,** and **ortho, ortho-substituted PCBs**, and they are also able to oxidise many of these compounds. The mechanism of induction is not well understood.

There is good evidence for the presence of cyt. P450 in earthworms. Thus **carbofuran** is metabolised to **3–hydroxycarbofuran** in *L. terrestris in vivo* (Stenersen *et al.* 1973). Pure parathion inhibits cholinesterase in *E. andrei in vivo*. This pesticide must be oxidised to **paraoxon** (diethyl 4–nitrophenyl phosphate) *in vitro* before being able to inhibit esterases. **Diethyl phosphate** is found amongst the metabolites of **parathion.** An oxidative and a hydrolytic step must therefore occur (Stenersen 1979b). However, **carbaryl** has a long-lasting sublethal toxicity in *E. andrei* (Stenersen 1979a), whereas this pesticide is rapidly hydroxylated, hydrolysed and conjugated in mammals (eg. Leeling and Casida 1966). Another interesting observation with relevance for the understanding of the cyt.-P450 system in the earthworm is that the toxicity of **2,3,7,8–tetrachlorodibenzoedioxin** seems to be very low (Reinecke and Nash 1984). This extremely potent mammalian toxicant acts by inducing cyt. P-450IA1 and certain other enzymes by binding to a protein (the Ah-receptor). The Ah-TCDD complex activates transcription of the genes involved (e.g. Whitlock 1989). Earthworms evidently lack this induction mechanism.

The presence of hemoblobin, with molecular weight similar to the microsomes, makes studies of cyt.-P450 difficult in earthworms. Liimitainen and

Hanninen (1982) solved this problem by using gel filtration chromatography in order to separate the interfering earthworm hemoglobin from the microsomes. They were thus able to detect cyt.- P450 in *L. terrestris* by direct spectrophotometric methods. Further studies were carried out by Milligan *et al.* (1986) who found cyt.-P450 at a concentration of 0·63–2·35 nmole/g gut tissue. They were unable to detect an increased amount of cyt.- P450 in *Dendrobaena veneta* after pretreatment with either **3–methylcholantrene** or **phenobarbitol** which are good inducers of the cyt-P450IA1 and cyt-P450II-types respectively in mammals. They concluded that noninducibility may be a general characteristic of earthworms. One would therefore expect them to accumulate nonpolar xenobiotics under conditions where constitutive levels of the enzyme are unable to metabolise the chemicals present in the body. On the other hand, the noninducibility may make substances that must be bioactivated by oxidation, such as certain carcinogens, less toxic. Berghout *et al.*(1990) found 0·51 nmol cyt.-P450 per g in the gut tissue of *L. terrestris*. They avoided the interference of hemoglobin by using the fluffy fraction overlaying the pellet, obtained by centrifugation at 100,000g. Benzoylresorufin, a model substrate of the 'bulky molecule'-type was oxidised at a V_{max} of 2,12 pmole/min. per mg protein, whereas ethoxyresorufin, a model substrate of the 'flat molecule'- type was not oxidised. These result may explain the low toxicity of **dioxin** found by Reinecke and Nash (1984). It should be mentioned that a marine polychete evidently has a cyt.P-450 that is induced by **polychlorinated biphenyls** (Lee *et al.* 1981), and more species should be studied before general conclusions on the role of cyt.-P450 in earthworms are made. However, it may be concluded that some xenobiotics are oxidised by the action of cyt.-P450 in earthworms although the main function of these enzymes may be in the oxidation of endogenous metabolites.

GLUTATHIONE TRANSFERASES

The glutathione transferases are present in all animals (Stenersen *et al.* 1987), and at least in many plants, microorganisms and fungi (Schrøder *et al.* 1990, Lau *et al.* 1980, Wackett and Gibson 1982). In earthworms they are present in all organs (Stenersen and Øien 1981). These enzymes constitute a family with many types which may be separated by isoelectric focusing (Stenersen *et al.* 1979), ion exchange chromatography (Stenersen and Øien 1981) and affinity chromatography and electrophoresis (Stenersen *et al.* 1987). The typical 'methyl-transferase' activity was not found in earthworms (Stenersen and Øien 1981) and they probably lack the ability to demethylate organophosphorus insecticides like **bromophos** and **parathion-methyl**. However, **quintozene** was conjugated by eight species out of eight tested *in vitro*. Khalil (1990) has shown that the glutathione conjugation pathway plays an important role in the metabolism of this fungicide *in vivo* in *Eisenia 'foetida'*. The ability to metabolise pesticides through the glutathione pathway is probably different in the various species because the conjugation of the model substrate **2,4–dichloronitrobenzene** is widely different (Stenersen and Øien 1981), as shown in Table 1.

Table 1. Activity of glutathione transferases in earthworms, showing the great species difference between activity towards 2,4-dinitrobenzene (CDNB) and 2,4-dichloronitrobenzene (DCNB) The data are from Stenersen and Øien (1981).

Species	Activity (CDNB) *nkat/mg protein	Activity (DCNB) % of CDNB-activity
A. rosea	35.9	0
A. caliginosa	7.5	0
A. longa	16.4	0.94
A. chlorotica	4.2	0
O. cyaneum	8.2	0
E. andrei	12.5	0 49
D. hortensis	3.2	0
L. rubellus	13.9	0.51
L. terrestris	10.5	0.23
E. albidus	6.7	0
R. norvegicus	24.1	2.1

* 1 nkat is the enzyme activity that conjugate 1 nmole substrate in 1 second.

OTHER CONJUGATION PATHWAYS

Other conjugation processes are evidently present in earthworms such as methylation, because substances such as tetrachloro-1,4–bis(methylthio) benzen are found as metabolites of **quintozene** (Khalil 1990). Formation of glucuronides, sulfates or other of the conjugates well-known from vertebrates have never been reported. Curl *et al.* (1987) have studied the metabolism of **cypermethrin** in *A. caliginosa* and *L. terrestris* and identified a spermine conjugate. This is an unusual product not reported in vertebrates. This finding demonstrates that further studies of pesticide metabolism in earthworms may give some new and interesting pathways.

HYDROLASES

Ester hydrolysis seem to be a pathway of some importance in earthworms. **Paraoxon** is easily hydrolysed, mainly by a microsomal hydrolase (Stenersen 1979b), and **cypermethrin** is hydrolysed *in vivo* (Curl *et al.* 1987). As pointed out by Roberts and Dorough (1985) some hydrolytic products are more toxic for earthworms than the parent compounds, even when these are potent pesticides, such as parathion. Hydrolysis is therefore not always a detoxication process.

BINDING

Earthworms seem to have an unusual ability to bind xenobiotics or their metabolites in unextractable forms. About half of the radioactivity originating from ^{14}C-ringlabelled **carbofuran** was bound in *L. terrestris* (Stenersen *et al.* 1973). There are some species differences in binding. Thus *Eisenia* excreted 95% of ^{14}C-**carbofuran** picked up, whereas *Lumbricus* retained 87%, mostly in unextractable form (Gilman and Vardanis 1974). Curl *et al.* (1987) who demonstrated a high accumulation rate of ^{14}C-**cypermethrin** in

A. caliginosa and *L. terrestris* (Fig. 4) showed that most of this material was bound in an unextractable form. The bound ^{14}C-residue did not accumulate in organisms eating earthworms. Haque and Ebing (1988) documented that a large amount of unextractables was formed from ^{14}C-**pentachlorophenol** in soil as well as in worms.

The identity and the toxicological significance of the bound residues that are frequently found in earthworms need further study.

Recommendations

Toxicity testing should be carried out in a way that eliminates the influence of soil adsorption because earthworms are potentially important as noncontroversial test organisms in toxicology. The artificial soil test is more relevant in practical ecotoxicology than the filter paper test, but tests should also be carried out to assess the toxicity without the complicating influence of the soil. Of the same reasons the metabolism of various pesticides and xenobiotics should be studied more intensively, possibly as a part of the pre-approval research on pesticides. Basic reasearch on the biotransformation of xenobiotics in earthworms are needed in order to assess their importance in the elimination process of these chemicals.

References

BERGHOUT, A., BÜLD, J. AND WENZEL, E. (1990). The cytochrome P-450-dependent monooxygenase system of the midgut of the earthworm *Lumbricus terrestris*. *Eur. J. Pharmacol.* **183** 1885–1886.

BEYER, W.N. AND KRYNITSKY, J. (1989). Long-Term persistence of dieldrin, DDT and heptachlor epoxide in earthworms *Ambio*, **18** 270–273.

CURL, E.A., EDWARDS, P.J., ELLIOTT, C. AND LEAHEY, J.P. (1987). Metabolites of cypermethrin in earthworms. *Pestic. Sci.* **20** 207–222.

GESTEL, V. C.A.M. AND MA, W.C. (1988). Toxicity and bioaccumulation of chlorophenols in earthworms in relation to bioavailability in soil. *Ecotox. Environ. Safety* **15** 289–297.

GILMAN, A.P. AND VARDANIS, A. (1974). Carbofuran. Comparative toxicity and metabolism in the worms *Lumbricus terrestris* L. and *Eisenia foetida* S. *J. Agr. Food Chem.* **22** 625–628.

GONZALES, F.J. (1990). Molecular genetics of the P-450 superfamily. *Pharmac. Ther.* **45** 1–38.

HAQUE, A. AND EBING, W. (1988). Uptake and accumulation of pentachlorophenol and sodium pentachlorophenate by earthworms from water and soil. *Sci. Total Environ.* **68** 113–125.

HEIMBACH, F. (1984). Correlation between three methods for determining the toxicity of chemicals to earthworms. *Pestic. Sci.* **15** 605–611.

KHALIL, M.A. (1990). Aufnahme und Metabolismus von ^{14}C-Hexachlorbenzol und ^{14}C-Pentachlornitrobenzol in Regenwrmer. *Dissertation, Techn. Univ. Mnchen, Germany* pp 137.

LAU, E.P., NISWANDER, L., WATSON, D. AND FALL, R.R. (1980). Glutathione S-transferase is present in a variety of microorganisms. *Chemosphere* **9** 565–569.

LEELING, N.C. AND CASIDA, J.E. (1966). Metabolites of carbaryl (1–naphthyl methylcarbamate) in mammals and enzymatic systems for their formation. *J.Agr. Food Chem.* **14** 281–289.

LIIMATAINEN, A. AND HÄÄNNINEN, O. (1982). Occurrence of cytochrome P-450 in earthworm *Lumbricus terrestris*. *In* Cytochrome P-450. Biochemistry, Biophysics and Environmenal implication (Edited by Hietanen E, Lartinen M and Hnninen) pp.255–258, Elsevier Amsterdam.

LORD, K.A., BRIGGS, G.G., NEALE, C.M., MANLOVE, R. (1980). Uptake of pesticides from water and soil by earthworms. *Pestic. Sci.* **11** 401–408.

MILLIGAN, L., BABISH, J., NEUHAUSER, F. (1986). Noninducibility of cytochrome P-450 in the earthworm *Dendrobaena veneta*. *Comp. Biochem. Physiol.* **85C** 85–87.

NEBERT, D.W. AND GONZALES, F.J. (1987). P-450 genes: structure, evolution and regulation. *Ann. Rev. Biochem.* **56** 945–993.

NEUHAUSER, E.F., LOEHR, R.C., MALECKI, M.R., MILLIGAN, D.C. AND DURKIN, P.R. (1985). The toxicity of selected chemicals to the earthworm *Eisenia fetida J. Environ. Qual.* **14** 383–388.

NEBERT, D.W., NELSON, D.R. AND FEYEREISEN (1989). Evaluation of the cytochrome P-450 genes. *Xenobiotica* **19** 1149–1160.

REINECKE, A.J. AND NASH, R.G. (1984). Toxicity of 2,3,7,8–TCDD and short-term bioaccumulation by earthworms (Oligochaeta). *Soil Biol. Biochem.* **16** 45–49.

ROBERTS, B.L. AND DOROUGH, W.H. (1985). Hazards of chemicals to earthworms. *Environ. Tox. Chem.* **4** 307–323.

SCHRØDER, P., LAMAOUREUX L.G. AND RENNEBERG, H. (1990). Glutathione S-transferase activity in spruce needles *Pestic. Biochem. Physiol.* **37** 211–218.

SOMASUNDARAM, L., COATS, J.R., RACKE, K.D. AND SHANBHAG, V.M. (1991). Mobility of pesticides and their hydrolysis metabolites in soil. *Environ. Tox. Chem.* **10** 185–195.

STENERSEN, J. (1979a). Action of pesticides on earthworms. Part I: The toxicity of cholinesterase-inhibiting insecticides to earthworms as evaluated by laboratory tests. *Pestic. Sci.* **10** 66–74.

STENERSEN, J. (1979b). Action of pesticides on earthworms. Part II: Elimination of parathion by the earthworm *Eisenia foetida* (Savigny). *Pestic. Sci.* **10** 104–112.

STENERSEN, J. (1984). Detoxication of xenobiotics by earthworms. *Comp. Biochem. Physiol.* **78C** 249–252.

STENERSEN, J., GILMAN, A. AND VARDANIS, A. (1973). Carbofuran: Its toxicity to and metabolism by earthworm (*L. terrestris*). *J. Agr. Food Chem.* **21** 166–171.

STENERSEN, J., GUTHENBERG C., AND MANNERVIK, B. (1979). Glutathione S-transferases in earthworms (*Lumbricidae*). *Biochem J.* **181** 47–50.

STENERSEN, J., KOBRO, S., BJERKE, M. AND AREND, U. (1987). Glutathione transferases in aquatic and terrestrial animals from nine phyla. *Comp. Biochem. Physiol.* **86C** 73–82.

STENERSEN, J. AND ØIEN, N. (1980). Action of pesticides on earthworms. Part IV: Uptake and elimination of oxamyl compared to carbofuran. *Pestic. Sci.* **11** 396–400.

STENERSEN, J. AND ØIEN, N. (1981). Glutathione S-transferases in earthworms (Lumbricidae). Substrate specificity, tissue and species distribution and molecular weight. *Comp. Biochem. Physiol.* **69C** 243–252.

WACKETT, L.P. AND GIBSON, T.D. (1982). Metabolism of xenobiotic compounds by enzymes in cell extracts of the fungus *Cunninghamella elegans*. *Biochem. J.* **205** 117–122.

WHITLOCK, J.P. JR (1989). The control of cytochrome P-450 gene expression by dioxin. *Trends in Pharmacol. Sci.* **10** 285–288.

Consequences for Vertebrate Wildlife of Toxic Residues in Earthworm Prey

A.S. COOKE[1], P.W. GREIG-SMITH[2] AND S.A. JONES[3]

[1] *English Nature, Northminster House, Peterborough, Cambs, UK*
[2] *Central Science Laboratory, MAFF, Tangley Place, Worplesdon, Surrey, UK*
[3] *Central Science Laboratory, MAFF, Hook Rise South, Tolworth, Surrey, UK*

Abstract

This paper reviews the potential risks to wildlife (chiefly birds and mammals) that may be caused by the effects of toxins in earthworms. The emphasis is on secondary poisoning by pesticides of predators that eat contaminated worms. Some attention is also given to other toxins (metals) and to indirect effects such as depletion of prey populations.

Introduction

Many species of wild vertebrates feed on earthworms, either as a regular, major part of their diet, or as an opportunistic strategy at times when earthworms are particularly abundant. There are two ways in which pesticides or other toxins that affect worms can have adverse consequences for their vertebrate predators.

First, there may be a risk of secondary poisoning through feeding on worms contaminated by toxic residues. In principle, this could occur whether or not the worms themselves suffer any adverse effects. However, for most chemicals the doses which are toxic to birds or mammals are usually also toxic to worms, and in poisoning incidents, dead worms are often found along with affected predators (eg Rudd, 1964; Bunyan *et al.*, 1981). Moreover, changes in the behaviour of earthworms after sublethal exposure (eg coming to the soil surface) will influence the likelihood of vertebrate exposure.

Second, depletion of earthworm populations in a habitat may have indirect consequences for certain species of birds and mammals by reducing their food supply, thereby leading to a lower predator population density without any obvious direct toxic effects.

This paper presents a brief review of some of the major issues in this topic, including: (1) an assessment of the bird and mammal species likely to be at risk from toxic residues in earthworms; (2) identification of the principal features of pesticides that may lead to secondary poisoning; (3) evidence for poisoning of vertebrates from field trials and incidents of wildlife mortality; (4) an assessment of the importance of indirect effects on earthworm predators caused by pesticide usage; (5) discussion about whether, and how, these potential risks should be taken into account in pesticide registration procedures.

The information considered is chiefly based on birds and mammals, and on UK agriculture. Although the emphasis is on pesticides, some data on the effects of other contaminants (metals) are included.

Species of birds and mammals at risk from pesticides

Macdonald (1983) has discussed much of the information available on the major earthworm predators, particularly the red fox (*Vulpes vulpes*) (see also Macdonald, 1980) and the badger (*Meles meles*) (Kruuk, 1978; Kruuk & Parish, 1981). These are species which specialise on earthworm prey, and organise their foraging behaviour according to the availability of worms, as do some birds, including gulls and plovers (Barnard & Thompson, 1985). A survey of information about diets, chiefly from Corbet & Harris (1991) and Cramp (1977–88), reveals a long list of species that occasionally or rarely eat

Table 1. Mammals and birds likely to be at risk in the UK from secondary poisoning by eating earthworms contaminated by agricultural pesticides.

Mammals	
Badger	*Meles meles*
Red fox	*Vulpes vulpes*
Mole	*Talpa europaea*
Hedgehog	*Erinaceus europaeus*
Common shrew	*Sorex araneus*
Wood mouse	*Apodemus sylvaticus*
Gulls and plovers	
Golden plover	*Pluvialis apricaria*
Lapwing	*Vanellus vanellus*
Black-headed gull	*Larus ridibundus*
Common gull	*Larus larus*
Herring gull	*Larus argentatus*
Predatory birds	
Little owl	*Athene noctua*
Tawny owl	*Strix aluco*
Buzzard	*Buteo buteo*
Kestrel	*Falco tinnunculus*
Game birds	
Pheasant	*Phasianus colchicus*
Woodcook	*Scolopax rusticola*
Thrushes	
Song thrush	*Turdus philomelos*
Blackbird	*Turdus merula*
Corvids	
Rook	*Corvus fugilegus*
Carrion crow	*Corvus corone*

earthworms, but worms are a major item of diet for only a small number. If species that do not feed in agricultural land (and are therefore unlikely to come into contact with poisoned worms) are excluded, the list is reduced further, to 15 species of birds and 6 species of mammals (Table 1). These form a suitable basis for assessment of the risks of poisoning, according to their respective predatory habits. It should be recognised that some species may be equally – or more – at risk from direct contact with pesticide granules, pellets etc than by eating the worms that are poisoned.

Features of pesticide usage liable to cause secondary poisoning

The chief uses of pesticides that may affect predatory vertebrates are seed treatments, sprayed and granular insecticides, and pelleted molluscicides. Several features are likely to increase the risks to predators from eating contaminated earthworms: (1) high toxicity; (2) high application rate, especially directly onto soil; (3) widespread use; (4) long persistence in soil and worms; (5) tendency to bioaccumulate.

The probability of such effects occurring may or may not be enhanced if the worms themselves are poisoned. A change in behaviour, or the occurrence of dead worms on the surface, may increase their likelihood of being taken by vertebrate predators or scavengers. On the other hand, if the worms do not die, they have longer to accumulate residues that persist in the soil.

The likelihood and severity of such effects may be influenced by the weather (some products cause problems in wet conditions – see later), the species of earthworms affected by the chemical, and their behaviour (eg activity at different depths in the soil), and the behaviour of the predators (eg specialisation, competition in flocks, density-dependent foraging tactics).

These factors can be used to construct a framework for exploring the relative risks posed by different products and types of formulation to the various categories of wildlife.

Uptake of pesticide residues by predators

The risks of toxic effects on vertebrates depend on the transfer of pesticide residues from earthworms. To model this process, it is necessary to know: (1) The distribution of pesticide residue concentrations in the population of earthworms, and changes after the pesticide is applied. This should be assessed for the various earthworm species, and for different parts of the soil profile, according to the foraging habits of relevant predators. (2) Any changes in earthworm behaviour that might influence the chances of predators taking them. These might increase or decrease exposure. (3) Doses causing both acute and chronic toxic effects on birds and mammals. (4) Intake of worms in relation to the body-size of the predator. This should take account of selection for or against particular species or size classes of earthworms (Barnard & Thompson, 1985), and other potential biases such as aversion by birds to the flavour of contaminated worms.

Information of this kind can be used to estimate the average probability of a bird or mammal predator ingesting a critical dose of a chemical by eating earthworms. If the data allow it, patterns of uptake may be described by sigmoid models (Brisbin *et al.*, 1986). This may be relevant to feeding on earthworms in a single session, in the case of a non-persistent chemical, or to cumulative intake in the case of a persistent or frequently-used compound.

Such assessments of individual risk can be extended to estimate the combined risk for the population of vertebrate predators at the site, by using data on variations in their exposure. For example, knowledge of the rates of feeding by dominant and subordinate birds, and their respective dietary specialisations, could allow an estimate of the proportion of the population that would exceed a toxic intake threshold.

This level of analysis is rarely attained. Most trials of pesticides in which the risks to worm-eating vertebrates are considered are limited to simple ratios between toxicity (measured by oral LD_{50} or LC_{50} tests in the laboratory) and residue levels in earthworms, measured in the field. By calculating the residue concentration of the compound in a number of earthworms that would provide a reasonable meal for a predator, it is possible to judge whether the likelihood of predators being poisoned is high or low, and the information also helps to interpret the cause of death of any casualties that are found.

Evidence for poisoning of vertebrates after feeding on contaminated earthworms

DDT

Much of the classic work on earthworm-bird food-chains was done in the United States where repeated local applications of DDT were used to control the vectors of Dutch elm disease (Rudd, 1964). American robins (*Turdus migratorius*) were killed, or their breeding success was impaired, to a degree that caused catastrophic declines in local populations. It appeared that earthworms accumulated significant residues of DDT (and DDE) from the soil via contaminated leaf litter (Table 2). Rudd (1964) stated that an

Table 2. Amounts of DDT and DDE present in soil, leaves and earthworms after the application of DDT to elm trees in summer. Residues were accumulated in earthworms, posing a potential risk to American robins (*Turdus migratorius*). Data from Rudd (1964).

	DDT	DDE
Application rate:	0.9 kg per tree	
Concentration of residues:		
soil surface	1·46 kg/ha	
top 5 cm of soil	5·9–9·5 ppm	0·7–5·4 ppm
leaves after spraying	174–273 ppm	9–20 ppm
leaves, after falling from tree	20–28 ppm	1–4 ppm
earthworms, soon after spraying	33–164 ppm	3–73 ppm
earthworms, several months later	86 ppm	33 ppm

INSECTICIDES - MECHANISM OF ACTION AND RESISTANCE

Edited by PROFESSOR Dr. D. OTTO and DR. B. WEBER (Germany).

Expected publication date: JULY 1992
approx. 405 pages; hardcover; index; 234x156mm,
ISBN 0 946707 38 3
Price £44.95

This book is a collection of over 30 papers written by experts reporting on the latest advances in research on:

a) Development of new concepts for ecologically orientated plant protection.

b) Latest applied research on commercial Insecticides and Pesticides.

c) Reports on worldwide research from leading specialists from 14 countries.

This book is based on fully edited and indexed contributions from the Symposium on "Insecticides - Mechanisms of Action and Resistance" held in Germany in April 1991.

Contributors are from:

Japan, Germany, Israel, U.K., Hungary, Switzerland, U.S.A., Cuba, Poland, U.S.S.R., China, Egypt and France.

Provisional contents Include:

- » Compounds which control developing hormonal processes of insects.
- » Regulation of the cockroach *Periplaneta americana.*
- » Selective insect control agents.
- » Insecticide that affects the insect cuticule.
- » Novel sulfur-containing compounds with juvenile hormone activity.
- » Species selective insect growth regulators.
- » Inhibition of mitochrondrial ATPase by its carbodiimide product.
- » Effects of a broad- spectrum antiviral agent on the hemimetabolous insect.
- » Plants as sources of novel pesticides.
- » Properties of Neemazal-F - a new botanical insecticide.
- » Three new insect growth inhibitors from *Prunella vulgaris.*
- » Antifeedants against stored product insects.
- » Proctolin and its physiological function in insects.
- » The GABA activated Cl channel in insects as target for insecticide action.
- » Effects of insecticides on the oxygen uptake of isolated cockroach ganglia.
- » Cotton pests resistance to pesticides and its management in the USSR.
- » The present status and contrameasures of insecticide resistance in vegetables in China.
- » Biochemical characterization of resistant insects.
- » Biochemical monitoring for insecticide resistance
- » Molecular biology of insecticide resistance.
- » Genetic start and end points of insecticide resistance evolution.
- » Trends in pesticide in resistance research.
- » Susceptibility of spider mite strains to amitraz and dicofol.
- » Metabolic resistance factors in *Tetranychus urticae.*

American robin eating 100 earthworms would ingest more than 3 mg DDT, a lethal dose. Many other bird species were affected similarly.

Rudd (1964) also described how forest spraying of only 0·6 kg DDT/ha to control spruce budworm (*Choristoneura fumiferana*) was associated with a significant reduction in numbers of young woodcock (*Scolopax rusticola*) in New Brunswick. It was suggested that woodcock acquired their DDT residues from contaminated earthworms. When the woodcock migrated to the southern United States they were contaminated by heptachlor sprayed at 1·7 kg/ha to control non-native fire ant infestations. Again it was suggested that the residues were derived from earthworms.

Dimond *et al.* (1970) studied DDT residues in soil, earthworms and American robins at different times after forests had been aerially sprayed with 1·1 kg/ha to control spruce budworm. Residues in robins were high in the year of treatment in part, it was suggested, because of direct contact with the spray. However, from year 2 to year 9, when residues in the robins were believed to have been derived from the food chain, no decline in residues could be detected. Residues in earthworms were relatively low, and although robins had more than 10 times the concentration in the worms, they showed no symptoms of poisoning.

When DDT was applied experimentally to a hay field, the time for disappearance of 50% of total DDT residues was 3·2 years, for both soil and earthworms (Beyer & Gish, 1980). A later paper about the same experiment by Beyer & Krynitsky (1989), which took account of earthworm residues over an even longer period, estimated the half-life of DDE to be 5·7 years. However, disappearance is not a simple process; it comprises several phases, occurring at different rates (Edwards, 1966). Also, persistence varies greatly according to soil type; Edwards (1966) reported a range from 26% to 78% DDT remaining after 3 years in a variety of soils. Beyer & Gish (1980) proposed that reported differences in accumulation rates by earthworms were due to a variety of factors, such as the feeding habits of the earthworms, soil type and sampling methodology. Davis (1971) demonstrated in laboratory trials a linear relationship between DDT soil residues and those in earthworms (see also Edwards & Lofty, 1972); Davis (1971) also showed an inverse relationship between residues in earthworms and organic matter content. Ploughing or rotovation sometime after spraying incorporates insecticide into the soil; this decreases the exposure of earthworms which feed at the surface, eg *Lumbricus terrestris*, but increases the exposure for subterranean species, eg *Allobophora caliginosa* (Davis & French, 1969).

Beyer & Gish (1980) pointed out that although the LC_{50}'s of DDT to four bird species in 5-day feeding trials ranged from 311 to 1869 ppm wet weight, poisoning of birds in the field was associated with residues in earthworms below this range. (Residues are quoted on a wet weight basis unless otherwise specified). These authors suggested a hazardous threshold level of 8 ppm. Earlier work by Barker (1958) indicated that residues of 30–40 ppm presented a lethal short-term hazard to bird predators of earthworms.

Poisoning of birds after eating DDT-contaminated earthworms has been noted in other countries, such as New Zealand, where blackbirds (*Turdus merula*) and song thrushes (*T. philomelos*) were poisoned (Collet & Harrison, 1968). In UK, there were several studies after DDT spraying in orchards (Bailey *et al.*, 1970, 1974; Moreton & Kite, 1975). As well as blackbirds and song thrushes, casualties of the following species were found poisoned by DDT: mistle thrush (*T. viscivorus*), grey partridge (*Perdix perdix*), red-legged partridge (*Alectoris rufa*), pheasant (*Phasianus colchicus*) and tawny owl (*Strix aluco*). In a Sussex orchard, studied by Bailey *et al.* (1970), earthworms were scarce and the residues in them were considered too low to have poisoned the birds directly. However, in an orchard in Cambridgeshire, Davis (1968) found high residues in worms, comparable to levels reported from North America. Davis' study site had been sprayed with 3·9 kg DDT/ha, compared with 1·4 kg/ha in the study of Bailey *et al.* (1970). Further work in an orchard in Norfolk sprayed with 2·1 kg DDT/ha (Bailey *et al.*, 1974) showed that thrushes could accumulate significant residues of DDT in their fat deposits from eating worms and other invertebrates. Deaths of birds were thought to occur because those residues were mobilised during times of reproductive stress. An alternative theory was that increased food intake during the breeding season may have been responsible. However, many gamebirds died soon after spraying, perhaps after grazing freshly-sprayed vegetation. In a Kent orchard studied by Moreton & Kite (1975), apples were sprayed with an annual average of 5·4 kg DDT/ha. Although blackbirds had reduced hatching and fledging success, the population level was unaffected because there was replacement by 'surplus' birds.

Moles (*Talpa europaea*) feed largely on earthworms (Gorman & Stone, 1990) and might be expected to have accumulated DDT residues in these orchard studies. Although moles caught in the Norfolk orchard had high residues of DDT compounds in their fat, they appeared to be unaffected and their liver residues were low (Bailey *et al.*, 1974). Rapid metabolism in the liver was suspected. Bailey *et al.* pointed out that contaminated moles could present a hazard to predatory birds feeding in orchard areas.

Although this paper concentrates on acute poisoning, DDT would more frequently have been passed onto birds via earthworm prey in quantities that cause subtle sub-lethal effects, particularly eggshell thinning (Cooke, 1973). The magnitude and frequency of such effects was often high enough to be translated into population declines (Newton, 1979).

Lower vertebrates may also be affected. Adult frogs, for instance, will consume earthworms (Frazer, 1983). The American bullfrog (*Rana catesbeiana*) is not sensitive to DDT, having a 96 hr LD_{50} of >2000 mg/kg (Hudson *et al.* 1984), but the Old World common frog (*R. temporaria*) is much more sensitive, with a 20-day LD_{50} of 7·6 mg/kg (Harri *et al.*, 1979). Based on data on food consumption and body mass given by Cooke (1974) for *R. temporaria* maintained under laboratory conditions, the content of DDT in their prey required to reach such an intake in a single day would be 1700–3300 ppm DDT. This gives a considerable safety margin compared with DDT residue levels in worms obtained in the field (up to 40 ppm in the UK; Davis, 1968). However, Cooke's frogs were fed with small food items, were relatively inactive and on average consumed less than 0·5% of their body mass per day. There is no reason why a 20 g frog should not consume a 2 g earthworm. In that case, residues in the worm of 75 ppm would be sufficient to produce an LD_{50} dose, yielding a considerably diminished safety factor.

If secondary poisoning by DDT occurred in *R. temporaria* in UK, it is more likely to have been caused by eating slugs than earthworms, because slugs are consumed more frequently (Cooke, 1974; Frazer, 1983) and contained at least comparable residues (Davis, 1968; Davis & French, 1969).

OTHER ORGANOCHLORINE INSECTICIDES

There have also been studies of uptake of other organochlorine insecticides by predators of earthworms. In laboratory tests, Davis (1971) showed that worms accumulated residues of dieldrin more rapidly than DDT. Earthworms in a potato field treated with aldrin were later found to contain both aldrin and dieldrin (Davis & French, 1969). The aldrin was thought to be on soil within the worms. Dieldrin residues in earthworms and other invertebrates in arable and orchard sites in Cambridgeshire were considered insufficient to cause acute lethal poisoning of birds (Davis, 1968). A captive song thrush was fed for six weeks on earthworms collected from an arable field previously sprayed with 3·4 kg aldrin/ha (Jefferies & Davis, 1968); the worms contained 0·3–0·6 ppm dieldrin, but the bird survived. The fate of other thrushes fed on earthworms experimentally-dosed with dieldrin indicated that the lethal exposure level was more than an order of magnitude higher. However, Beyer & Gish (1980) found that plots treated with 2·2 kg dieldrin/ha could contain residues in worms, of a magnitude hazardous to thrushes, for three years. Comparing the residues in earthworms found by Davis (1968) and Beyer & Gish (1980), it appears that aldrin spraying resulted in lower levels of dieldrin in earthworms than did dieldrin spraying. The principal use of dieldrin in Davis' study area was as a seed dressing; it was applied at only 0·14–0·17 kg/ha and produced fairly low residues in the worms. Residues considered to be hazardous were found in earthworms after spraying plots with 2·2 kg heptachlor expoxide/ha (Beyer & Gish, 1980).

A poisoning incident that may have involved earthworms and an organochlorine pesticide was recorded in Nottinghamshire in 1987 (Greig-Smith *et al.*, 1989). Twelve thrushes were found with several organochlorine residues such as might be obtained from an approved chlordane-based earthworm treatment.

Earthworms are eaten by badgers (*Meles meles*) in large numbers, and deaths of badgers from dieldrin poisoning were evidently quite common in eastern England during the 1960s (Jefferies, 1969). However, Jefferies (1969) attributed the mortality to eating poisoned woodpigeons (*Columba palumbus*) rather than earthworms.

Contamination of soil with 2, 3, 7, 8-tetrachlorodibenzo-p-dioxin could give rise to secondary poisoning of mammals, as indicated by data summarised by Verschueren (1983). Earthworms collected after the contamination incident at Seveso, Italy in 1979 contained up to 0·012 μg/g. LD_{50}'s for laboratory mammals are as low as 0·6 μg/kg (guinea pig). A small mammal consuming an earthworm diet contaminated at 0·012 μg/g would be expected to have a daily intake exceeding the LD_{50}.

It is relevant to ask why DDT caused secondary poisoning incidents but none was reported for dieldrin. Davis (1968) considered that soil invertebrates could provide a regular chronic source of organochlorine residues for ground- feeding birds, but that levels were too low to cause acute poisoning. However, Davis did not study worm residues from any field sprayed earlier in the same year with dieldrin, although within his study area two carrot fields were sprayed with 0·8 and 1·2 kg dieldrin/ha. The field plots of Beyer & Gish (1980), that were sprayed with 0·6 kg/ha, produced residues in earthworms of about 8 ppm dry weight, equivalent to approximately 2 ppm wet weight. In a laboratory study, Davis (1971) found that worms maintained in soil treated at a rate equivalent to 3 kg dieldrin/ha, accumulated up to 4·6 ppm in four weeks. Jefferies & Davis (1968) reported that a single song thrush maintained on worms containing 6 ppm survived, but another maintained on a dietary level of 12 ppm died. Because wild birds died when exposed to DDT residues at which captive birds of other species survived (Beyer & Gish, 1980), it seems reasonable to suggest that dieldrin may have caused some mortality in birds that consumed contaminated earthworms. However, in view of the major incidents caused by dieldrin used as a seed dressing it is not surprising that such incidents seem to have been overlooked. The rapid disappearance of soft-bodied earthworms from the guts of affected birds will not have helped in this respect. Had there been a cue from the United States, as there was with DDT, dieldrin incidents involving worms could have been looked for specifically in the UK, but dieldrin was not used as extensively in north America (Newton, 1979).

The properties that probably account for problems with DDT are its extreme persistence and ability to bioaccumulate, together with moderate toxicity. Dieldrin is generally more toxic to vertebrates in terms of acute poisoning (eg Hudson *et al.*, 1984). DDT also had widespread use, often at quite high rates of application.

OTHER PESTICIDES

Aldicarb is a soil-applied systemic carbamate insecticide, used as granules since the mid 1970s to protect potatoes, sugar beet and onions from attack by insects and nematodes. In 1975 and 1976 a series of incidents was investigated through the UK Wildlife Incident Investigation Scheme (Anon, 1978; Hardy *et al.*, 1986). Casualties numbered up to 100 birds per incident and the species affected included the black-headed gull (*Larus ridibundus*), the common gull (*L. canus*), the lapwing (*Vanellus vanellus*), the moorhen (*Gallinula chloropus*) and gamebirds. Incidents were characterised by granules being applied to wet soil and left exposed on the soil surface before rotovation. It was suggested that birds such as gulls were poisoned either because they had ingested the aldicarb granules directly or had taken contaminated earthworms. An intensive field study was carried out by Bunyan *et al.* (1981) to investigate the environmental consequences of aldicarb seed treatment for sugarbeet. Although dead earthworms were found on the soil surface 6 days after drilling, they did not contain detectable residues of aldicarb. Later surveillance on other fields revealed moribund earthworms, containing aldicarb residues, on the soil surface in wet conditions. When the reason for the incidents was understood, modifications were made to the application machinery to minimise the number of granules on the surface. The attention of users was drawn to this hazard. The only incidents since were in 1984 and 1988, when poor incorporation allowed granules to remain on the surface of fields and gulls and partridges were found dead (Fletcher & Hardy, 1984; Greig-Smith *et al.*, 1990). Earthworms were present in the gizzards of the gulls examined.

Phorate is a systemic granular organophosphorus insecticide used against a range of pests on many crops. Five incidents of bird mortality occurred between 1983 and 1988 (Fletcher & Hardy, 1984, 1985; Fletcher *et al.*, 1989; Greig-Smith *et al.*, 1990). Lapwings, pheasants, red-legged partridges, common gulls and black-headed gulls were killed. Dead earthworms were seen on the soil surface in at least one of the incidents (Fletcher & Hardy, 1984) and were probably the route by which the birds were poisoned.

The chemical control of moles is practised by using strychnine (eg Gorman & Stone, 1990). Earthworms, preferably *Lumbricus terrestris*, are treated with strychnine and placed individually in deep burrows. Only about 25 are needed, perhaps with several baitings, to clear one hectare, indicating the high potential for residues in earthworms to kill vertebrate predators.

In 1986 in Devon a buzzard (*Buteo buteo*) was found dead with earthworms in its beak (Fletcher *et al.*, 1989). Although mole control with strychnine had been carried out at the site, residues of carbofuran were confirmed in the bird. Carbofuran is a systemic carbamate insecticide and nematicide for soil treatment used to protect a range of crops and nursery stock. Carbofuran has been implicated in a number of other wildlife incidents in Britain (eg Fletcher & Hardy, 1983, 1984; Fletcher *et al.*, 1989; Greig-Smith *et al.*, 1990). Several of these incidents involved wet field conditions and some dry. In at least one incident, birds were thought to have eaten the granules (Fletcher & Hardy, 1983); in none of the incidents were earthworms mentioned. However, carbofuran has been observed to kill earthworms in the United States (see Smith, 1987). Moreover, amongst the birds poisoned in the British incidents were two little owls (*Athene noctua*), which would not be expected to take carbofuran granules directly, but do feed readily on earthworms (Cramp, 1985), and therefore poisoning from earthworm prey must be suspected. Soft-bodied prey such as earthworms might be overlooked during investigations and their importance may well be under-estimated.

The properties of the organophosphate and carbamate insecticides that appear to have caused such effects on wildlife are very different to the organochlorines. Aldicarb, carbofuran and phorate are all acutely toxic to higher vertebrates (Smith, 1987). In the UK, these active ingredients are commercially available only in granular formulations (Ivens, 1990), which makes them safer to apply for the human operator. It also makes them safer for the environment generally, as there is no risk of spray drift. However,

Table 3. Information on selected organophosphate and carbamate insecticides. The first group have been implicated in British incidents in which earthworms poisoned birds; the second group have not. Information sources: Hudson *et al.*, 1984; Smith, 1987; Worthing & Walker, 1987.

Active ingredient	LD_{50} (mg/kg)*		Persistence in soil	Bioaccumulation potential
	Quail spp	Duck spp		
Aldicarb	2·0–4·7(5)	3·4	Moderate	Low
Carbofuran	5·0–12(3)	0·2–0·5(4)	Moderate	Low
Phorate	7·0–21(2)	0·6–2·6(2)	Moderate	Low
Carbosulfan	82	8·1	Low	?
Disulfoton	12–29(2)	6·5	Moderate	Low
Oxamyl	4·2–9·4(3)	2·6	Low	?

* Range with number of tests in brackets

Table 4. Potential hazards to birds from insecticide/molluscicide products, based on the ratio of application rates and toxicity. The number of LD_{50}s per m^2 provides an approximate 'hazard rating' for pesticide uses.

Active ingredient/ use	Application rate (kg/ha of a.i.)	Species	LD_{50} mg/kg	LD_{50}s (per kg bird) per m^2
Aldicarb/granular insecticide	1–4	Typical sensitive bird	2·0	50–200
Carbofuran/granular insecticide	0·34–5	Typical sensitive bird	0·2	170–2500
Phorate/granular insecticide	1·1–3·9	Typical sensitive bird	0·6	183–650
Dieldrin/ seed dressing	0·44	Pigeon	27	1·6
Triazophos/ foliar spray insecticide	0·14–0·42	Japanese quail	4–27	0·5–10·5
Chlorfenvinphos/ seed dressing	0·76–4·5	Pigeon	16	4·8–28
Carbophenothion/ seed dressing	0·24	Anser goose spp	29–35	0·7–0·8
Metaldehyde/ molluscicide broadcast as pellets or admixed	0·45–0·9	Ducks/ geese	0·3–0·8	56–300
Methiocarb/ molluscicide broadcast as pellets or admixed	0·22	Typical sensitive birds	4·6–13	1·7–4·8

they are still available for uptake by soil/surface-feeding organisms, and all three are toxic to earthworms (Edwards & Lofty, 1972). Their toxicity to selected bird species is shown in Table 3. For the most sensitive bird species, the application rate is equivalent to approximately 100 LD_{50}'s (for a 1 kg bird) per m^2 (Table 4). Thus, for a 1 kg duck, an LD_{50} dose is available in each 100 cm^2 of soil surface. In contrast, their potential to bioaccumulate is low, but they are moderately persistent in soil with half-lives of up to a few months (Smith, 1987). Clearly, they present a considerable hazard to birds through direct consumption of granules and it is not surprising to find they may also cause secondary poisoning via earthworms.

We now examine and compare other organophosphate and carbamate insecticides which are also commercially available in the UK only in granular formulations. There are three active ingredients for which comparable bird toxicity data have been published: carbosulfan, disulfoton and oxamyl (Table 3). Of these, carbosulfan appears less acutely toxic and no wildlife incidents involving carbosulfan have been reported in the UK. Oxamyl, although of comparable toxicity to aldicarb, is of low persistence in soil (Smith, 1987), and no wildlife incidents have been reported. Disulfoton appears to be of lower toxicity than aldicarb, carbofuran or phorate, but is moderately persistent (Smith, 1987). Disulfoton caused four known wildlife incidents in the UK during the period 1980–2, but none appeared to involve contaminated earthworms (Hardy *et al.*, 1986). In the United States, carbofuran and phorate have caused incidents and there was an unconfirmed incident implicating disulfoton (Smith 1987); earthworms were not thought to be involved in any incident.

Curl *et al.* (1987) assessed the potential for bioaccumulation of cypermethrin, a synthetic pyrethroid insecticide, in quail and rats fed on contaminated earthworms. Their conclusions from laboratory experiments were that vertebrates do not further accumulate the conjugated residues that are taken up by earthworms. The low application rates of pyrethroid products ensure that residues in the field are unlikely to reach levels that would be toxic to birds and mammals.

METALS

It is also worth briefly considering uptake of toxic metals by earthworms in contaminated environments. Worms can accumulate mercury (including methylmercury) from the soil (Bull *et al.*, 1977; Beyer *et al.*, 1985). There is no evidence that vertebrate wildlife has been affected by mercury accumulated from earthworms in agricultural situations, although mercury contamination of wildlife is of general concern (eg Cooke *et al.*, 1982; Beyer *et al.*, 1985). Also, there was a series of major incidents on the Mersey estuary 1979–1981, when several thousand wildfowl, waders and gulls died from alkyl lead poisoning after eating contaminated invertebrates, including polychaete worms (Bull *et al.*, 1983). It is relevant to note that organolead compounds have similar properties to the organochlorines, being persistent, able to bioaccumulate and toxic (Newton *et al.*, 1980; Verschueren, 1983). They might therefore be expected to cause problems if wildlife is exposed to residues in earthworms.

There has long been concern about the amount of cadmium reaching agricultural soils from deposited sewage sludge (eg Royal Commission on Environmental Pollution, 1979). Earthworms are known to accumulate cadmium from soil (Ireland & Richards, 1981), which in the case of sewage sludge reaches levels that are potentially dangerous to predators (Beyer *et al.*, 1982). Little owls are predators of earthworms (see above); Newton *et al.* (1984) analysed cadmium in the livers of 23 little owls found dead in various parts of England, 1981–3. Only six contained more than 1 ppm dry weight, up to a maximum of 6 ppm. The authors concluded that this did not suggest widespread or heavy contamination of little owls.

Indirect effects of pesticides caused by depletion of earthworm populations

Pesticides or other contaminants may affect birds and mammals indirectly by reducing the density of earthworm prey on arable land or grass. Depending on local conditions, such changes in the available food supply may reduce the density of predators (Tucker, in press), or may cause them to shift their foraging effort elsewhere (Barnard & Thompson, 1985; Macdonald, 1980).

Earthworm densities are very low in some soils (Edwards, 1983), and the effects of pesticides on earthworms are unlikely to be important for vertebrates in such areas, even though the impact on earthworm populations may be proportionately greater than in high density areas.

Long term reduction of earthworm populations by pesticides depends as much on the persistence and frequency of use of pesticides as on their toxicity. It is likely to occur where molluscicides or granular insecticides are used regularly, such as on crops of potatoes and sugarbeet (Davis *et al.*, 1990; Sly, 1986). These are not habitats that are heavily exploited by foraging birds, although it is difficult to determine whether low densities of earthworms are a contributory factor to the scarcity of birds.

At a national scale, surveys of bird population trends in the UK have indicated some patterns consistent with widespread reductions of invertebrate populations on farmland, after the introduction of soil insecticides during the 1970's (Marchant *et al.*, 1990).

An additional indirect effect may be caused by changes in habitat that can arise as a consequence of lowered earthworm activity, and hence reduced fertility, in certain soils, affecting the growth and species-composition of vegetation. Edwards & Lofty (1972) have pointed out that soil fertility, and hence crop performance, is dependent on earthworm activity. This implies that farmers could benefit by abstaining from the use of pesticides that would adversely affect earthworm populations.

Registration tests: when to request them and what to request

What are the implications for registration data requirements if it is suspected that a new pesticide may cause vertebrate mortality because of consumption of contaminated worms? It seems that extreme persistence and potential to bioaccumulate led to incidents being caused by the moderately toxic organochlorine insecticides. New compounds with such properties are unlikely to be proposed for registration and are virtually certain not to be approved.

Although the organophosphate and carbamate insecticides have low potential to bioaccumulate, they are moderately persistent in soil with half-lives up to a few months. The compounds that cause problems to earthworms and vertebrates are formulated as granules and applied at approximately 100 LD_{50}'s (per 1 kg bird) per m^2. Such application rates should be sufficient to cause registration authorities automatic concern over safety of vertebrate wildlife. Other proven poisoning cases are usually associated with much lower hazard ratings (Table 4). The mode of exposure in these situations is different (ie pesticide residues are on vegetation, soil or seed rather than incorporated as granules) and the pesticide may therefore be more or less available for uptake by worms. However, there has been no evidence that contaminated worms were the cause of these other incidents. Therefore, a suitable trigger for being concerned about worms seems to be when the application of a granular formulation approaches or exceeds 100 LD_{50}'s (per 1 kg bird) per m^2.

In such cases, information on toxicity to worms in the laboratory should be supplied as part of the initial main package of data submitted to the registration authority. Wildlife trials should be requested, as it will be clear from bird toxicity data that feeding on the granules will almost certainly lead to mortality in the field. Information on worms should be gathered in these field trials. In relation to bird and mammal poisoning, this will need to include: (1) a quantitative assessment of pesticide residues in worms over time; (2) information on changes in the behaviour of worms, dead worms on the surface, or birds being attracted to dead worms; (3) consideration should also be given to the artificial flooding of plots on the study fields, as wildlife incidents involving earthworms have sometimes been associated with flooded conditions. If there is an interest in the worms *per se*, then population assessments should be made to determine whether earthworm densities decline.

Thus information on worms in relation to bird mortality will not be requested as a separate test but as part of a wildlife trial that will need to be undertaken anyway. It may also be necessary to ask for a laboratory feeding trial with another species of bird, in addition to the standard duck and quail. Tests on gulls may be both realistic and feasible, but would represent a departure from the usual range of species tested.

With modern toxic granular formulations, bird poisoning after eating earthworms may be less important than direct poisoning after eating granules. However, in order to be able to undertake remedial action with any confidence, it is necessary to understand potential routes and mechanisms of poisoning. Therefore the relative importance of the poisoning route via earthworms must be studied and understood.

Conclusions

It is clear that the introduction of certain pesticides in the UK has been accompanied by incidents of wildlife mortality attributable to feeding on earthworms containing toxic residues. This is particularly so for soil-applied granular insecticides, where some evident risks have been alleviated by adjustment of the conditions of use in the light of early experience. However, these examples are only the most obvious cases. Arrangements for identifying adverse effects due to secondary poisoning are imperfect, because of the limitations of investigations of wildlife incidents to detect mortality, and especially sublethal effects, on some species (Greig-Smith, in press). Also, it is rarely easy to establish the relative risks of poisoning through feeding on worms and on the pesticide formulation itself.

Although a number of mammal species feature prominently in Table 1 as potential victims of secondary poisoning after eating contaminated worms, it is generally birds that are recorded as casualties in incidents. When deciding whether a particular level of wildlife mortality is acceptable, the 'conservation value' of the species involved is important. Cooke (1990) divided British birds and mammals at risk from pesticide poisoning into 'common' and 'key' species. Some key species were classified because of legal protection rather than their rarity. Of the birds in Table 1, only the golden plover (*Pluvialis apricaria*) and the predatory birds are key species. Other key species referred to in this paper are the mistle thrush and the grey partridge. So far, however, numbers of key species killed have been very low.

Confidence in the safety of pesticide products can be enhanced by paying more attention to possible risks due to secondary poisoning in the design of field trials, as outlined in the previous section. Also, certain properties (persistence, bioaccumulation) should be considered alongside information on toxicity and environmental fate, to identify circumstances of potential hazard, by means of explicit risk estimation, as proposed by Rothert & Kokta (1992) and Greig-Smith (1992).

Labelling of pesticide products can also help to reduce the risks of poisoning. In the UK, although information about the toxicity of new pesticides to worms is gathered at the registration stage, precautionary phrases are not included on the product labels. The inclusion of phrases such as 'harmful to earthworms' or 'dangerous to earthworms' would provide a user with additional guidance to help in choosing the most environmentally appropriate pesticide, minimising both direct and indirect effects on vertebrate wildlife.

References

Anon (1978). Review of wildlife incidents. Pest Infestation Control Laboratory Triennial Report 1974–76. MAFF, London.

BAILEY, S., BUNYAN, P.J., JENNINGS, D.M., TAYLOR, A. (1970). Hazards to wildlife from the use of DDT in orchards. *Pesticide Science* **1**, 66–69.

BAILEY, S., BUNYAN, P.J., JENNINGS, D.M., NORRIS, J.D., STANLEY, P.I., WILLIAMS, J.H. (1974). Hazards to wildlife from the use of DDT in orchards: II. A further study. *Agro-Ecosystems* **1**, 323–338.

BARKER, R.J. (1958). Notes on some ecological effects of DDT sprayed on elms. *Journal of Wildlife Management* **22**, 269–274.

BARNARD, C.J., THOMPSON, D.B.A. (1985). *Gulls and Plovers. The Ecology and Behaviour of Mixed-Species Feeding Groups*. London, Croom Helm.

BEYER, W.N., CHANEY, R.L., MULHERN, B.M. (1982). Heavy metal concentrations in earthworms from soil amended with sewage sludge. *Journal of Environmental Quality* **11**, 381–385.

BEYER, W.N., CROMARTIE, E., MOMENT, G.B. (1985). Accumulation of methylmercury in the earthworm, *Eisenia foetida*, and its effect on regeneration. *Bulletin of Environmental Contamination and Toxicology* **35**, 157–162.

BEYER, W.N., GISH, C.D. (1980). Persistence in earthworms and potential hazards to birds of soil applied DDT, dieldrin and heptachlor. *Journal of Applied Ecology* **17**, 295–307.

BEYER, W.N., KRYNITSKY, A.J. (1989). Long-term persistence of dieldrin, DDT, and heptachlor epoxide in earthworms. *Ambio* **18**, 271–273.

BRISBIN, I.L., WHITE, G.C., BUSH, P.B. (1986). PCB intake and the growth of waterfowl: multivariate analyses based on a reparameterized Richards sigmoid model. *Growth* **50**, 1–11.

BULL, K.R., EVERY, W.J., FREESTONE, P., HALL, J.R., OSBORN, D., COOKE, A.S., STOWE, T. (1983). Alkyl lead pollution and bird mortalities on the Mersey Estuary, UK, 1979–1981. *Environmental Pollution (series A)* **31**, 239–259.

BULL, K.R., ROBERTS, R.D., INSKIP, M.J., GOODMAN, G.T. (1977). Mercury concentrations in soil, grass, earthworms and small mammals near an industrial emission source. *Environmental Pollution* **12**, 135–140.

BUNYAN, P.J., VAN DEN HEUVEL, M.J., STANLEY, P.I., WRIGHT, E.N. (1981). An intensive field trial and a multi-site surveillance exercise on the use of aldicarb to investigate methods for the assessment of possible environmental hazards presented by new pesticides. *Agro Ecosystems* **7**, 239–262.

COLLETT, N., HARRISON, D.L. (1968). Some observations on the effects of using organochlorine sprays in an orchard. *New Zealand Journal of Science* **11**, 371–379.

COOKE, A.S. (1973). Shell thinning in avian eggs by environmental pollutants. *Environmental Pollution* **4**, 85–152.

COOKE, A.S. (1974). The effects of pp^1-DDT on adult frogs (*Rana temporaria*). *British Journal of Herpetology* **5**, 390–396.

COOKE, A.S. (1990). The acceptability of pesticide-induced damage to wildlife. In: *Pesticide Effects on Terrestrial Wildlife*, L. Somerville & C.H. Walker (Eds). Taylor & Francis, London. pp. 291–306.

COOKE, A.S., BELL, A.A., HAAS, M.B. (1982). *Predatory Birds, Pesticides and Pollution*. Institute of Terrestrial Ecology, Cambridge.

CORBET, G.B., HARRIS, S. (1991). *The Handbook of British Mammals*. 3rd Edition. Blackwell Scientific Publications, Oxford.

CRAMP, S. (Ed) (1977–1988). *Handbook of the Birds of Europe, the Middle East and North Africa*. Vols I–V, Oxford University Press, Oxford.

CURL, E.A., EDWARDS, P.J., ELLIOTT, C., LEAHEY, J.P. (1987). The conjugation and accumulation of metabolites of cypermethrin by earthworms. *Pesticide Science* **20**, 207–222.

DAVIS, B.N.K. (1968). The soil macrofauna and organochlorine insecticide residues at twelve agricultural sites near Huntingdon. *Annals of Applied Biology* **61**, 29–45.

DAVIS, B.N.K. (1971). Laboratory studies on the uptake of dieldrin and DDT by earthworms. *Soil Biology and Biochemistry* **3**, 221–233.

DAVIS, B.N.K., FRENCH, M.C. (1969). The accumulation and loss of organochlorine insecticide residues by beetles, worms and slugs in sprayed fields. *Soil Biology and Biochemistry* **1**, 45–55.

DAVIS, R.P., GARTHWAITE, D.G., THOMAS, M.R. (1990). Arable farm crops 1988. Survey Report 78. MAFF Publications, London.

DIMOND, J.B., BELYEA, G.Y., KADUNCE, R.E., GETCHELL, A.S., BLEASE, J.A. (1970). DDT residues in robins and earthworms associated with contaminated forest soils. *Canadian Entomologist* **102**, 1122–1130.

EDWARDS, C.A. (1966). Insecticide residues in soils. *Residue Reviews* **13**, 83–132.

EDWARDS, C.A., LOFTY, J.R. (1972). *Biology of Earthworms*. London, Chapman & Hall. 283 pp.

FERGUSON, D.E., GILBERT, C.C. (1967). Tolerances of three species of anuran amphibians to five chlorinated hydrocarbon insecticides. *Journal of the Missouri Academy of Science* **13**, 135–138.

FLETCHER, M.R., HARDY, A.R. (1983). Pesticides and wildlife: a review of wildlife incidents investigated from October 1981 to September 1982. MAFF Reference Book 252, pp. 47–55.

FLETCHER, M.R., HARDY, A.R. (1984). The impact of pesticides on wildlife: a review of the wildlife incidents investigated from October 1983 to September 1984. Reference Book No. 257 (84). MAFF, London.

FLETCHER, M.R., HARDY, A.R. (1985). The impact of pesticides on wildlife: a review of the wildlife incidents investigated from October 1984 to September 1985. Reference Book No. 257 (85). MAFF, London.

FLETCHER, M.R., HUNTER, K., GREIG-SMITH, P.W., RUTHVEN, A.D. (1989). Investigations of suspected poisoning of animals by pesticides in Great Britain, October 1985 to December 1986. Report of the Environmental Panel of the Advisory Committee on Pesticides. MAFF, London.

FRAZER, D. (1983). *Reptiles and Amphibians in Britain*. Collins, London.

GORMAN, M.L., STONE, R.D. (1990). *The Natural History of Moles*. Christopher Helm, London.

GREIG-SMITH, P.W. (in press). Understanding the impact of pesticides on wild birds by monitoring incidents of poisoning. In: *The Population Ecology and Wildlife Toxicology of Agricultural Pesticide Use: A Modelling Initiative for Avian Species*, R.J. Kendall & T. Lacher (Eds).

GREIG-SMITH, P.W. (1992). Risk assessment approaches in the UK for the side-effects of pesticides on earthworms. In *Ecotoxicology of Earthworms* (H. Becker *et al*, eds) pp. 159–168. Intercept Ltd, Hants, UK.

GREIG-SMITH, P.W., FLETCHER, M.R., HUNTER, K., QUICK, M.P., RUTHVEN, A.D., SHAW, I.C. (1989). Investigations of suspected poisoning of animals by pesticides in Great Britain in 1987. Report of the Environmental Panel of the Advisory Committee on Pesticides. MAFF, London.

GREIG-SMITH, P.W., FLETCHER, M.R., HUNTER, K., QUICK, M.P., RUTHVEN, A.D., SHAW, I.C. (1990). Pesticide poisoning of animals 1988: investigations of suspected incidents in Great Britain. Report of the Environmental Panel of the Advisory Committee on Pesticides. MAFF, London.

HARDY, A.R., FLETCHER, M.R., STANLEY, P.I. (1986). Pesticides and wildlife: twenty years of vertebrate wildlife incident investigations by MAFF. *State Veterinary Journal* **40**, 182–192.

HARRI, M.N.E., LAITINEN, J., VALKAMA, E.-L. (1979). Toxicity and retention of DDT in adult frogs, *Rana temporaria* L. *Environmental Pollution* **20**, 45–56.

HUDSON, R.H., TUCKER, R.K., HAEGELE, M.A. (1984). Handbook of toxicity of pesticides to wildlife. US Dept. of the Interior, Fish & Wildlife Service, Washington DC.

IRELAND, M.P., RICHARDS, K.S. (1981). Metal content, after exposure to cadmium, of two species of earthworms of known differing calcium metabolic activity. *Environmental Pollution (Series A)* **26**, 69–78.

IVENS, G.W. (Ed) (1990). *The UK Pesticide Guide*. CAB International/British Crop Protection Council, Farnham.

JEFFERIES, D.J. (1969). Causes of badger mortality in eastern counties of England. *Journal of Zoology*, London, **157**. 429–436.

JEFFERIES, D.J., DAVIS, B.N.K. (1968). Dynamics of dieldrin in soil, earthworms, and song thrushes. *Journal of Wildlife Management* **32**, 441–456.

KRUUK, H. (1978). Foraging and spatial organisation of the European badger, *Meles meles* L. *Behavioral Ecology and Sociobiology* **4**, 75–89.

KRUUK, H., PARISH, T. (1981). Feeding specialisations of the European badger, *Meles meles* in Scotland. *Journal of Animal Ecology* **50**, 773–788.

MACDONALD, D.W. (1980). The red fox, *Vulpes vulpes*, as a predator upon earthworms, *Lumbricus terrestris*. *Zeitschrift fur Tierpsychologie* **52**, 171–200.

MACDONALD, D.W. (1983). PREDATION ON EARTHWORMS BY TERRESTRIAL VERTEBRATES. IN: *Earthworm Ecology*. J.E. Satchell (Ed.) London, Chapman and Hall. pp. 393–414.

MARCHANT, J.H., HUDSON, R., CARTER, S.P., WHITTINGTON, P. (1990). *Population Trends in British Breeding Birds*. Tring, Herts, British Trust for Ornithology.

MORETON, B.D., KITE, T.E. (1975). DDT and birds in a Kentish orchard. *Bird Study* **22**, 228–232.

NEWTON, I. (1979). *Population Ecology of Raptors* T. & A.D. Poyser, Berkhamsted.

NEWTON, I., OSBORN, D., BELL, A.A., BULL, K.R. (1980). *Birds of Prey and Pollution*. Nature Conservancy Council Contract Report No. 303 Institute of Terrestrial Ecology, Monks Wood Experimental Station, Huntingdon.

NEWTON, I., BELL, A.A., BULL, K.R., FREESTONE, P., GORE, D.J., HAAS, M.B., HALL, J.R., HANSON, H.M., HARRIS, M.P., LEACH, D.V., OSBORN, D., WYLLIE, I., YOUNG, W.J. (1984). *Birds and Pollution*. Nature Conservancy Council Contract Report No. 543 Institute of Terrestrial Ecology, Monks Wood Experimental Station, Huntingdon.

ROTHERT, H., KOKTA, C. (1992). Hazard and risk assessment scheme for evaluating effects of pesticides on earthworms. In *Ecotoxicology of Earthworms*, Intercept Ltd, Hants, UK, pp. 169–176.

Royal Commission on Environmental Pollution (1979). *Seventh Report, Agriculture and Pollution*. HMSO, London.

RUDD, R.L. (1964). *Pesticides and the Living Landscape*. Faber & Faber, London.

SLY, J.M.A. (1986). Review of usage of pesticides in agriculture, horticulture and animal husbandry in England and Wales 1980–83. Survey Report 41. HMSO, London.

SMITH, G.J. (1987). Pesticide use and toxicology in relation to wildlife: organophosphorus and carbamate compounds. Resource Publication 170. US Dept. of the Interior, Fish & Wildlife Service, Washington DC.

TUCKER, G.M. (in press). Effects of agricultural practices on field use by invertebrate – feeding birds in winter. *Journal of Applied Ecology*.

VERSCHUEREN, K. (1983). *Handbook of Environmental Data on Organic Chemicals*. 2nd edition. Van Nostrand Reinhold, New York.

WORTHING, C.R., WALKER, S.B. (Eds) (1987). *The Pesticide Manual*. British Crop Protection Council, Thornton Heath.

Application of hazard and risk assessment schemes for earthworms

Risk Assessment Approaches in the UK for the Side-effects of Pesticides on Earthworms

P.W. GREIG-SMITH

Ministry of Agriculture, Fisheries and Food, Central Science Laboratory, Tangley Place, Worplesdon, Surrey GU3 3LQ, UK

Abstract

In the UK arrangements for the registration of pesticides, effects on earthworms are considered chiefly because of the potential risks of secondary poisoning for birds and mammals. The decision-making procedures used will be formalised, and extended to examine the risks of population changes that may affect soil fertility, through an initiative of the European Plant Protection Organization and the Council of Europe. As part of a broader development of environmental risk assessment schemes, this will address the particular aspects of earthworm ecotoxicology that are relevant to pesticide regulation.

Introduction

In the UK the approval of pesticides for agricultural uses is covered by the Food and Environment Protection Act 1985 (FEPA), which operates through the Control of Pesticide Regulations (COPR), implemented in 1986. Within this legislation there are provisions for ensuring the safety of products to operators, consumers and the environment. These are based on assessments made by the Ministry of Agriculture, Fisheries and Food (MAFF) on behalf of six government departments, within a framework agreed by, and under the constant overview of an independent expert body, the Advisory Committee on Pesticides (ACP) (see MAFF/HSE, 1991).

These arrangements provide the basis for a detailed prior assessment of any likely environmental problems that might arise from new products or different uses of existing ones. In addition, post-registration surveillance is provided by the national Wildlife Incident Investigation Scheme, under which suspected incidents of poisoning of wildlife, livestock, companion animals and honeybees are examined (Greig-Smith, 1988).

Although broadly similar arrangements have been in operation for many years, the range of environmental side-effects considered in registration is continually expanding. In part, this indicates the growth of knowledge about the effects of pesticides, and improvements in the techniques available to study them, but it also reflects an increasing concern about long-term and cumulative impacts on animal populations.

Toxic effects of pesticides on earthworms have been considered chiefly when a potential risk of secondary poisoning for worm-eating birds and mammals is identified. In such cases, initial calculations of likely exposure are based on the application rate of the product, its frequency of use, and the potential for bioaccumulation of the active ingredient. This may be supplemented by a data requirement for measurement of residues in earthworms in the course of field trials (see Hardy, 1990; Cooke, Greig-Smith & Jones, 1992).

The need for such data is determined through a flexible, stepwise approach in which expert judgement is applied on a case by case basis. While this has been proved in the past to achieve satisfactory standards of environmental safety, a more formalised decision-making process that still permits flexibility and expert judgement would have many advantages. Also, the conclusions that result from current assessments are based on poorly-defined criteria which do not fully take account of the secondary poisoning risk (see later). Further, they fail to address other possible consequences of earthworm contamination. Accordingly, there is recognition that a more detailed, and wider-ranging assessment procedure is desirable. It is likely to be provided through the activities of an expert panel on environmental risk assessment established jointly by the European Plant Protection Organization (EPPO) and the Council of Europe (Smith, 1990). This body is developing regulatory decision-making schemes for a broad range of environmental issues, including the effects of pesticides on earthworms.

This paper briefly reviews the present role of earthworm data in UK registration of pesticides, and describes how it may change as a result of the EPPO/Council of Europe initiative and other developments, to give a better assessment of environmental consequences.

Relevance of effects on earthworms

There are three reasons why adverse effects of pesticides on earthworms may be important. First, worms play a major role in the maintenance of fertility in most soils (Edwards & Lofty, 1977). This means that the use of products which reduce their numbers or activity may have consequences for crop performance that counteract their intended plant protection function. Similarly, vegetation in field margins and neighbouring land may be affected, in ways that jeopardise the habitats of beneficial animals, both vertebrate and invertebrate.

Second, reductions in the density of worms will reduce the quality of the habitat for their predators, who may switch to forage elsewhere, or may suffer impaired breeding success.

Third, the presence of toxic residues in earthworms poses a risk of secondary poisoning for predators. Indeed, tertiary effects are possible, for example if a bird of prey captures a small rodent that has eaten contaminated worms. Such longer food-chains are important if the chemical involved has bioaccumulation properties.

Only the last of these categories is usually considered in registration. In principle, risks to predators could be judged empirically by residue measurements in field trials. However, routes of exposure are often more complex (eg birds may eat both worms and pesticide granules) and a better understanding of mechanisms is desirable (see Cooke *et al.*, 1992).

The first two types of side-effect require a different approach, focusing on reductions in worm density. This entails criteria based on the *size* of population changes, and the time for *recovery*, which may vary greatly from chemical to chemical and from species to species (Jepson, 1989, Greig-Smith, 1992). There may also be changes in adult: juvenile ratios, or the relative numbers of different species. Unacceptable levels of change or its duration must also take account of 'normal' fluctuations in numbers (cf. Domsch *et al.*, 1983).

Thus there are two distinct kinds of risk assessment required. One deals with the probability that worm density will fall to critical levels or for critical periods, adversely affecting crop performance, food supplies for predators, and habitat (each of which will have a different criterion). The other concerns the probability that toxic residues in worms will accumulate to levels that cause poisoning of predators.

Present UK requirements

Official information on the data requirements for UK registration is given in MAFF (1986); this is intended as initial guidance that is neither comprehensive nor up-to-date, and is frequently supplemented by additional requirements appropriate to the particular uses of a given product. Nevertheless, the information required on earthworms is generally similar. It involves an initial LC_{50} test using artificial soil (OECD, 1984), which may be supplemented by observations made in the course of field trials with broader aims, to evaluate short-term risks of poisoning of birds and mammals. Samples of worms for residue analysis are obtained from soil by irrigating with dilute formaldehyde (Raw, 1959), by digging, or by an electrical method which brings worms to the surface (Thielemann 1986). Counting and collection of dead worms found on the surface is also valuable, as a means of assessing the availability of, and toxic burdens present in, worms affected by a pesticide application (eg Bunyan *et al.*, 1981). This can reveal the influence of weather conditions on the toxic effects (eg Fig. 1).

MAFF (1986) also indicates methods to be used for measuring changes in earthworm populations, based on quadrat sampling from replicated plots that should include a toxic standard (eg chlordane or benomyl) as well as the test compound (Edwards & Brown, 1982). However, such tests are rarely required at present.

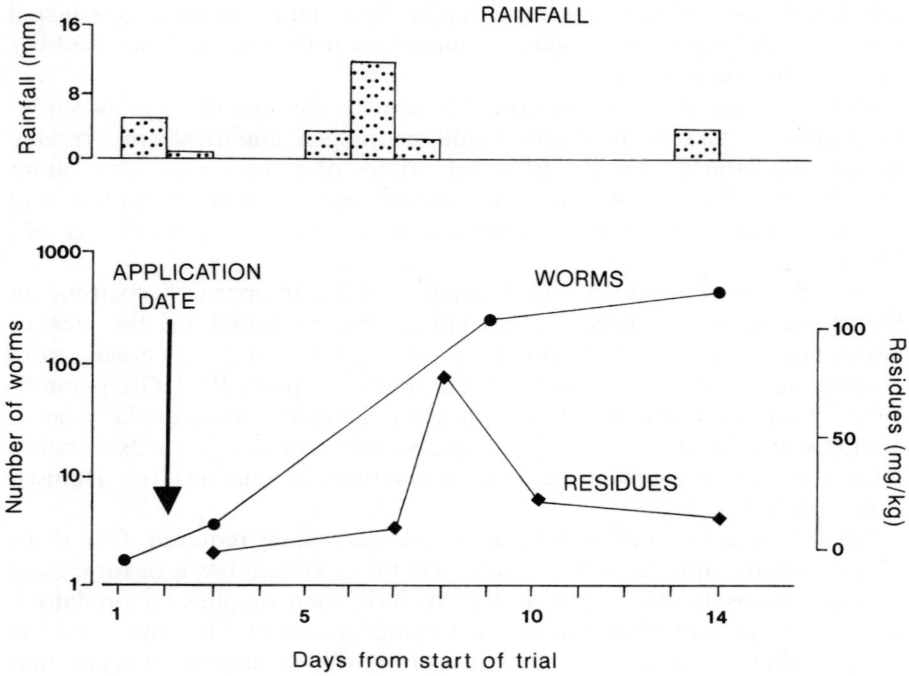

Figure 1. Example of data collected in a field trial of a soil-applied insecticide, showing an increase in the numbers of worms found dead on the surface of the field, and the residue levels in them, after rainfall.

The EPPO/Council of Europe approach

The principles behind the decision-making schemes being developed by the EPPO/Council of Europe panel are outlined by Greig-Smith (1991b). Briefly, the aim is to provide a framework to guide regulators on the issues that need to be addressed, the data required, and how they should be interpreted. This is achieved by a stepwise approach that can be adapted to a very detailed, complex pathway if the data allow it and regulatory bodies desire it, or can be abbreviated to a simpler, unrefined assessment. One important feature is that the decision-making chain should not replace expert judgement with automatic standard 'triggers' to determine a further tier of testing or the classification of products. Rather, the scheme should identify where such judgement is essential, and include it within the procedure.

The framework comprises a series of stages that culminate in an *estimation* of risk, which is then followed by *classification* of the product (eg. as being of high, medium or low risk) and risk *management*, specifying conditions of approval designed to minimise the risks in practice (Fig. 2). Assessment proceeds from basic data on the proposed uses and properties of pesticide, to the environmental fate of the active ingredient (distribution and persistence), its toxicity to the species likely to be at risk, and the levels of

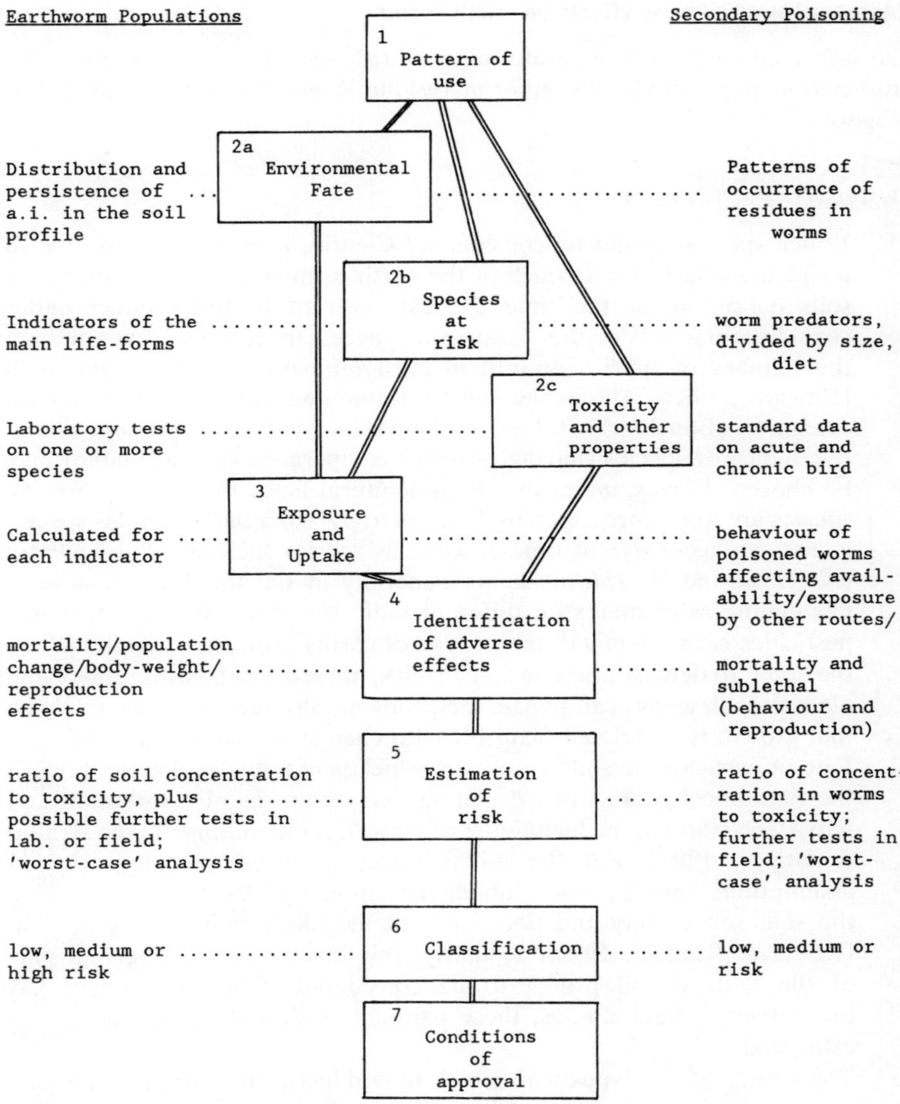

Figure 2. Framework for the assessment of the environmental risks caused by pesticide use (from Greig-Smith, 1991b), showing how the approach can be adapted to assess effects on earthworm populations and the risks of secondary poisoning for predators of earthworms.

exposure experienced under typical conditions of use. In many of these stages, there are particular issues raised by effects on earthworms and consequent secondary poisoning problems.

Assessment of adverse effects on earthworms

As indicated above, two separate routes of risk assessment are required, one directed at populations, the other at residue levels. These are indicated in Figure 2.

POPULATION CHANGES

1. Which species should be considered? Clearly, it would be desirable to adapt the assessment to each of the earthworm species occurring in the soils concerned at the time of year relevant to the product under consideration. It is neither feasible nor necessary to do so, even though the number of species present in most temperate soils is quite small (Edwards, 1983). This range can be accommodated in a few ecological categories (Bouché, 1977, Lavelle, 1983), for each of which a representative indicator species (having previous comparative data available) may be chosen. For example, in UK agricultural land, the deep-burrowing species are well represented by *Lumbricus terrestris* (although this species is also surface-active at times), whereas species such as *Allolobophora caliginosa* and *A. chlorotica* are generally in the top few cm of soil. Combining assessments on different indicators may be difficult, if two pesticides each affect a different type of earthworm, for example. When the aim is to detect effects on soil fertility, more direct measures involving plant growth assays can bypass the problem, although it is wise to check that growth is correlated plausibly with changes in worm numbers.
2. Fate of pesticides in soil is a subject which is rapidly developing, both in the theoretical aspects of calculating likely patterns of distribution and persistence, and in the techniques of empirical measurement (see Arnold & Briggs, 1990). For the initial stages of assessment, a series of assumptions must be made, about the amount of the product reaching the soil, soil density and porosity, and the likely rates of degradation (Kokta & Rothert, 1992). Similarly, this involves specifying that part of the vertical soil profile to be considered. Once assessment has progressed to field studies, these parameters should be measured, not estimated.
3. The choice of soil types with which to conduct tests is of great importance, in view of the major effects this may have on earthworm populations (Edwards & Lofty, 1977). For laboratory tests, the merits of various substrates have been widely debated (Heimbach, 1984). As part of the EC harmonization of pesticide registration procedures (Thomas, 1990), studies of pesticide fate in soils are becoming standardised to cover a range of representative 'European soils'. While this usefully promotes consistency of results obtained by different laboratories, and allows

comparison of different active ingredients it carries dangers of missing effects that may be expressed in local conditions of use. This can be partially accommodated by employing safety factors and ensuring that the assessment involves a 'worst-case' analysis (see below).

4. For some purposes, the numbers of worms may not be a sufficient index of pesticide effects. If their activity or reproduction is affected, the functional consequences may be as serious as those caused by mortality. Therefore, a series of sublethal end-points to tests may be appropriate, (eg Bouwman & Reinecke, 1987) if the properties of the chemical suggest such effects, perhaps through comparison with previous experiences of similar products.
5. Pesticide effects are bound to be variable, because of differences in local soils, drainage, earthworm populations, weather conditions at the time of use, and deviations from good agricultural practice. To deal with this, risk assessments should take account of extreme, 'worst-case' conditions, as well as the typical, average conditions (cf. Hart, 1990). It is not appropriate to attempt to simulate such extremes in all field studies, although that may sometimes be possible. A better approach is to extrapolate beyond the normal range of conditions, to predict what might happen under different circumstances. This is inevitably speculative, but can be supported by the occurrence of incidents revealed by monitoring a limited area of commercial use (Greig-Smith, 1988). That is the only means of reliably detecting the influence of extreme weather conditions, although if specific mechanisms are suspected, artificial simulations in field trials may be worthwhile. For example, the effect of rain or flooding can be easily reproduced by watering experimental plots.
6. A key part of most assessment schemes is to compare the environmental availability of a chemical with its toxicity to animals (eg. Rothert & Kokta, 1992). This is done mainly to determine whether further studies are necessary, but may also form the basis for classifying products once all the necessary information has been obtained. The safety factors used to give these criteria are arbitrary, and are generally established by reference to existing products, or by choosing a very large margin of safety, avoiding the need for detailed knowledge. However this is done, the sensitivity of risk estimates to variations in these thresholds should be incorporated into the assessment procedure.

SECONDARY POISONING

7. The species of predators likely to occur where a pesticide is used can be divided into a small number of groups, each represented by one or more indicators. This should take account of feeding behaviour that might affect exposure (eg. species that pick worms from the surface, versus those that probe into soil), differences in susceptibility to toxic effects and dietary habits (eg specialist worm-feeders versus opportunists). Separate assessments may be necessary for each group.

8. As for worm populations, it is useful to calculate the ratio between the availability to predators of residues in worms and their toxicity, as a means of guiding further studies. However, this cannot easily be achieved without empirical measurement of residue distributions, because of the many untested assumptions that would be involved in a theoretical calculation. There are other complications in determining the exposure of predators to residues. Many species are liable to feed on both the pesticide formulation itself (granules, pellets, sprayed vegetation) and on poisoned earthworms (eg Johnson *et al.*, 1991). In such cases, it is difficult to distinguish between the two routes of exposure, and preliminary assessment based on the application rate of the product (eg. the 'number of lethal doses per square metre', Cooke *et al.*, 1992) is usually sufficient. Effects on the behaviour of worms which influence the chances of predation must also be taken into account. For example, feeding birds may specialise on dead worms on the soil surface, or conversely they might develop aversions to poisoned worms through distaste or conditioned aversion learning.
9. One particular issue relates to bioaccumulation of residues in worms. Predators may be at a high risk if frequent use of persistent chemicals leads to a build-up of residues in worms, especially if there is a longer predatory food-chain involving amphibians or small rodents, for example. Bioaccumulation in earthworms may be measured directly (Beyer *et al.*, 1985; Curl *et al.*, 1987), but a rough indication can be obtained from the physical properties of the active ingredient. If the octanol-water partition coefficient is high, bioaccumulation is potentially likely (Lord *et al.*, 1980) and should be taken into account in the assessment of risk.

Conclusion

The principles upon which the proposed risk assessment scheme depend allow for a stepwise progression from initial desk calculations, to laboratory tests, and to field studies, when necessary. However, these stages are closely allied to other parts of the environmental assessment of pesticides. Thus, the effects on earthworm populations depend on knowledge of distribution and persistence in soil, while the levels of residues taken up, and the effects these have on worms, become an integral part of assessing the risks to predatory wildlife. At least in the field, studies of earthworms are always likely to be carried out as part of larger, multidisciplinary trials addressing a wide range of environmental side effects (Greig-Smith, 1990).

Acknowledgements

I am grateful to Carol Aldridge for comments on the manuscript.

References

ARNOLD, D.J., BRIGGS, G.G. (1990). Fate of pesticides in soil: predictive and practical aspects. In: *Environmental Fate of Pesticides* D.H. Hutson & T.R. Roberts (Eds), Chichester, John Wiley & Sons. pp. 101–122.

BEYER, W.N., CROMARTIE, E.; MOMENT, G.B. (1985). Accumulation of methylmercury in the earthworm, *Eisenia foetida*, and its effect on regeneration. *Bull Environ. Contam. Toxicol.* **35**, 157–162.

BOUCHÉ, M.B. (1977). Strategies lombriciennes. In: *Soil organisms as components of ecosystems*. U. Lohm and T. Persson (Eds), *Proc. 6th Int. Soil Zool Coll., Ecol. Bull. (Stockholm)* **25,** 122–132.

BOUWMAN, H., REINECKE, A.J. (1987). Effects of carbofuran on the earthworm *Eisenia fetida*, using a defined medium. *Bull. Environ. Contam. Toxicol.* **38**, 171–178.

BUNYAN, P.J, VAN DEN HEUVEL, M.J., STANLEY, P.I., WRIGHT, E.N. (1981). An intensive field trial and a multi-site surveillance exercise on the use of aldicarb to investigate methods for the assessment of possible environmental hazards presented by new pesticides. *Agro-Ecosystems* **7**, 239–262.

COOKE, A.S., GREIG-SMITH, P.W., JONES, S.A (1992). Consequences for vertebrate wildlife of toxic residues in their earthworm prey. In *Ecotoxicology of Earthworms* (H. Becker *et al.*, eds). pp. 139–155 Intercept Ltd, Hants, UK.

CURL, E.A., EDWARDS, P.J., ELLIOTT, C., LEAHEY, J.P. (1987). The conjugation and accumulation of metabolites of cypermethrin by earthworms. *Pesticide Science* **20**, 207–222.

DOMSCH, K.H., JAGNOW, G., ANDERSON, T.-H. (1983). An ecological concept for the assessment of side-effects of agrochemicals on soil microorganisms. *Residue Reviews* **86**, 65–105.

EDWARDS, C.A. (1983). Earthworm ecology in cultivated soils. In: *Earthworm Ecology* J.E. Satchell (Ed), London, Chapman and Hall, pp. 123–137.

EDWARDS, C.A., LOFTY, J.R. (1977). *Biology of Earthworms* 2nd edition. London, Chapman and Hall, 333pp.

EDWARDS, P.J., BROWN, S.M. (1982). Use of grassland plots to study the effect of pesticides on earthworms. *Pedobiologia* **24**, 145–150.

GREIG-SMITH, P.W. (1988). Wildlife hazards from the use, misuse and abuse of pesticides. *Aspects of Applied Biology* **17**, 247–256.

GREIG-SMITH, P.W. (1990). Intensive study versus extensive monitoring in pesticide field trials. In: *Pesticide Effects on Terrestrial Wildlife* L. Somerville and C.H. Walker (Eds), London, Taylor and Francis, pp. 217–239.

GREIG-SMITH, P.W. (1991a). The Boxworth experience : effects of pesticides on the fauna and flora of cereal fields. In: *The Ecology of Temperate Cereal Fields* L.G. Firbank, N. Carter, J.F. Darbyshire and G.R. Potts (Eds), Oxford, Blackwell Scientific Publications, pp. 333–371.

GREIG-SMITH, P.W. (1991b) Environmental risk assessment of plant protection products: an approach to the development of guidelines. *Bull. OEPP/EPPO Bull.* **21**, 219–226.

HARDY, A.R. (1990). Estimating exposure: the identification of species at risk and routes of exposure. In: *Pesticide Effects on Terrestrial Wildlife* L Somerville and C.H. Walker (Eds), London, Taylor and Francis, pp. 81–97.

HART, A.D.M. (1990). The assessment of pesticide hazards to birds: the problem of variable effects. *Ibis* **132**, 192–204.

HEIMBACH, F. (1984). Correlation between three methods for determining the toxicity of chemicals to earthworms. *Pesticide Science* **15**, 605–611.

JEPSON, P.C. (1989). The temporal and spatial dynamics of pesticide side- effects on non-target invertebrates. In: *Pesticides and Non-target Invertebrates* P.C. Jepson (Ed), Wimborne, Intercept, pp. 95–127.

JOHNSON, I.P., FLOWERDEW, J.R., HARE, R. (1991). Effects of broadcasting and of drilling methiocarb molluscicide pellets on field populations of wood mice, *Apodemus sylvaticus*. *Bull. Environ. Contam. Toxicol.* **46**, 84–91.

KOKTA, C., ROTHERT, H. (1992). Hazard and risk assessment for effects of pesticides on earthworms – the approach in the Federal Republic of Germany. In *Ecotoxicology of Earthworms* (H. Becker *et al.*, eds). pp. 169–176 Intercept Ltd, Hants, UK.

LAVELLE, P. (1983). The structure of earthworm communities. In: *Earthworm Ecology* J.E. Satchell (Ed), London Chapman and Hall, pp. 449–466.

LORD, K.A., BRIGGS, G.G., NEALE, M.C., MANLOVE, R. (1980). Uptake of pesticides from water and soil by earthworms. *Pesticide Science* **11**, 401–408.

MAFF (1986). Data requirements for approval under the Control of Pesticides Regulations 1986. London, MAFF.

MAFF/HSE (1991). *Pesticides 1991*. Reference Book 500, London, HMSO, 426 pp.

OECD (1984). Guideline for testing of chemicals no. 207. Earthworm, acute toxicity tests. Organisation for Economic Cooperation and Development, Paris.

RAW, R. (1959). Estimating earthworm populations by using formalin. *Nature (London)* **184**, 1661–1662.

SMITH, I.M. (1990). EPPO and the harmonization of registration requirements. In: *Future changes in pesticide registration within the EC* B. Thomas (Ed), Farnham, BCPC, pp. 87–90.

THIELEMANN, U. (1986). Elektrischer Regenwurmfang mit dert Oktett-Merthode. *Pedobiologia* **29**, 297–303.

THOMAS, B. (Ed) (1990). *Future changes in Pesticide Registration within the EC*. Monograph No. 44. British Crop Protection Council, Farnham.

A Hazard and Risk Assessment Scheme for Evaluating Effects of Pesticides on Earthworms: The Approach in the Federal Republic of Germany

C. KOKTA AND H. ROTHERT

Federal Biological Research Centre for Agriculture and Forestry, Department for Plant Protection Products and Application Techniques, Zoology Division, Messeweg 11/12, D-3300 Braunschweig, Germany

Abstract

Risk assessment is based on estimation of exposure and evaluation of adverse effects. A sequential testing procedure is established starting with an acute test on earthworm toxicity in the laboratory. The toxicity is compared with the estimated environmental concentration. Results of this comparison, data on environmental fate and the frequency of application are triggers for further testing. Assessment of risk to earthworms is followed by risk management such as labelling with recommendations for safe use or reducing the number of applications.

Introduction

In the Federal Republic of Germany the evaluation of side effects on non-target organisms is part of the clearance procedure for plant protection products. As stated in Article 15 of the 1986 Plant Protection Act ('Pflanzenschutzgesetz') registration will be granted when – besides other criteria – the plant protection product, when used for its intended purpose and in the correct manner 'does not have any other effects, particularly with regard to the balance of nature, which are not justifiable in the light of the present state of scientific knowledge'.

Earthworms play a major role in improving and maintaining soil fertility (Edwards & Lofty, 1977) and they are an important part of terrestrial food webs. Therefore possible effects of pesticides on earthworms have to be investigated prior to registration (Bode *et al.*, 1988). Because of their important role a recent proposal of the EEC ('Council directive concerning the placing of accepted plant protection products on the market', in press) also requires earthworm testing.

Test methods and testing procedure

SEQUENTIAL TESTING PROCEDURE

In order to assess effects of pesticides on earthworms, tests with environmental relevance are needed. To identify chemicals that may have critical effects and study them in detail, a stepwise testing procedure has proved to be useful, for example in evaluating effects on honeybees, and seems to be useful in earthworm testing.

All pesticides that will be used outdoors and can reach the soil have to be tested for possible effects on earthworms. For other pesticides, for example such as for protection of storage products, no data are required. Herbicides that are applied locally to single plants, or rodenticides that are placed into burrows or in bait stations are also excluded from testing.

For pesticides that can reach the soil, acute toxicity has to be determined as the first step of the testing procedure. Existing international guidelines (OECD, 1984; EEC, 1985) are appropriate for testing acute toxicity. Besides mortality, this test also provides some information on body weight development of adult earthworms. Additional information on the fate of the pesticide in soil, its degradation products and the estimation of exposure help to determine whether a test on sublethal effects such as reproduction and body weight development should be done as the second step. There are several proposed guidelines for a laboratory test on reproduction of *Eisenia fetida*, for example by Van Gestel *et al.* (1989) and Kula & Kokta (in press). Testing reproductive toxicity using species of arable land does not look very promising at present as these species have a generation time of several months. It also seems to be difficult to standardize test conditions.

Field testing is intended to be the third step within the testing procedure. Pesticides which are hazardous in the laboratory should be tested under practical conditions in the field. Edwards & Brown (1982) and Kula (in prep.) have proposed designs for field tests on grassland plots.

SELECTION OF TEST SPECIES

Availability of an appropriate test species for laboratory testing is a very important prerequisite for the development of assessment schemes. As there are only about ten common earthworm species in the northern parts of Europe (Dunger, 1983) choice might not be as difficult as in other taxonomic groups of soil organisms with several hundreds of species. According to Bouché (1977), there are three main ecological groups of earthworms. The deep-burrowing species live in permanent burrows and take food from the soil surface. The surface-active species live in the litter layer and the mineral-soil-dwellers inhabit the mineral soil. It is appropriate to test a representative of each of these types. This is realised in field tests on grassland sites because species of all these ecological categories normally are present.

However, there are problems in testing representatives of these ecological categories in the laboratory. Although Lofs-Holmin (1982) and Martin (1986) presented methods for laboratory breeding of species of arable land, especially *Aporrectodea caliginosa*, these are still difficult to rear in sufficient numbers in the laboratory. Instead, *Eisenia fetida* has proved to be a suitable laboratory test organism because of its high reproductive potential and the ease of captive breeding. In central Europe, however, the occurrence of this species is restricted to places such as dung heaps that are rich in organic matter.

Several authors have compared the sensitivity towards pesticides of *Eisenia fetida* and other species (Haque & Ebing 1983; Heimbach 1985; Kula, in press). Studies like these can help to establish a factor to take account of uncertainty in risk assessment when using *Eisenia fetida* instead of species of arable land. According to the present knowledge a factor of about 10 seems to be appropriate in most cases when transferring results from *Eisenia fetida* to other species.

In special cases it might be necessary to undertake toxicity testing of the degradation products of a pesticide.

Exposure analysis

There are two main pathways by which worms can be exposed to pesticides. The first is by contact with the body surface. Ingestion of contaminated soil or food is the other important route of exposure. According to Van Gestel & Ma (1990) the toxicity of a pesticide is primarily determined by the soil-pore-water concentration.

In order to assess the risk for earthworms, the pesticide's toxicity has to be compared with the concentration which is expected to occur initially in the field (EEC: estimated environmental concentration). Modelling of earthworm exposure is difficult as it depends on the behaviour of both earthworms and the pesticide. Therefore estimation of the initial amount of pesticide on the soil surface and in the soil is based on the following assumptions:

1. *The pesticide is evenly distributed in the upper 2·5 cm of the soil*. This means that exposure estimation is similar for all ecological groups of earthworms. The depth of 2·5 cm is an average value for the activity zones of different species and may lead to an overestimation of exposure of the mineral-soil-dwellers and an underestimation of exposure of the surface-active deep burrowing species and surface-living species. As juvenile earthworms live in the upper few centimeters of the soil, their exposure might also be suitably assessed by the depth of 2·5 cm.
2. *The average bulk density of the soil is 1·5 g/cm^3*. A bulk density of 1·4 to 1·5 g/cm^3 is typical for agricultural soil and of 1·5 to 1·6 g/cm^3 for soils with permanent culture. Therefore a mean bulk density of 1·5 g/cm^3 seems appropriate. Extreme soils with a low bulk density are not

normally under agricultural use and should therefore not be used in field tests.

3. *If the pesticide is used on bare soil or on soil with little plant cover the exposure estimation is based on the total amount applied*. Most herbicides applied on arable land are used for weed control before or immediately after sowing. At this time there is little crop cover, and nearly all of the amount applied will reach the soil. On grassland, the full amount applied will reach the grass layer but does not necessarily reach the soil. Stringer & Lyons (1974) found that after spraying orchards nearly all of the pesticide was on the grass layer but only a small amount directly on the soil. Nevertheless, it is assumed that the full amount is relevant for earthworms, because the grass layer is a potential food source for the deep-burrowing *Lumbricus terrestris* and other *Lumbricus* species living in the grass layer.
4. *If the pesticides is applied on land with a substantial plant cover it is assumed that 50% will reach the soil surface*. Previous studies suggest that an average of 50% is appropriate. For example Shires (1985) found that 60–75% of the amount applied in a spring wheat field at ear emergence reached the soil surface. Stringer & Lyons (1974) found that 80% of the amount applied in a fruit garden at an early bud stage in April and 50% at full canopy in June reached the soil. It is obvious that there will be some cases where the amount available on soil is much lower than 50% (for example in a cereal field during flowering). If data on residues in soil are available they should be used for the exposure estimation.
5. *For pesticides with less than three applications in a season the total amount will be added up, making allowance for its persistence in soil*. If a pesticide is persistent (e.g. DT50 › 60 days, triggers concerning earthworms are not fixed yet) the full amount of three applications should be added up. If the pesticide is less persistent, the addition should be done by counting only 50% of the second and the third application.

This procedure for estimation of exposure is valid for liquid applications only. The procedure in the case of granules is not yet decided. Perhaps it is not possible to do an overall estimation and testing should incorporate relevant exposure pathways from the beginning.

Hazard and risk assessment scheme

As a first step of the assessment procedure a comparison of the LC_{50} (Lethal Concentration for 50% of test animals) and the EEC is carried out. If the LC_{50} exceeds the EEC by a factor of 100 or more, the pesticide is classified as having low risk for earthworms (Fig. 1).

The factor of 100 includes several components to provide a margin of safety in the assessment:

1. In some cases *Eisenia fetida* is expected to be less susceptible to pesticides than species of arable land (Haque & Ebing, 1983).

2. There might be differences in sensitivity of the different life stages of earthworms.
3. The standardized test substrate in the laboratory ('artificial soil') has a higher content of organic matter than natural soils. Nevertheless, Van Gestel & Ma (1990) found for some chemicals that adsorption onto artificial soil is similar to that onto natural soils.
4. Mortality is the predominant effect taken into account when doing risk estimation at this stage, and sublethal effects are not considered.

If the LC_{50} is lower than 10 × EEC, use of the pesticide is assumed to involve a potential risk for earthworms. Further testing is required to obtain information on chronic effects and to determine whether these effects may occur in practical use.

If the LC_{50} is between 10 × EEC and 100 × EEC, effects on earthworms might occur, depending on circumstances. Additional criteria, such as the mobility and persistence of the pesticide in soil, the application pattern and other factors must be used to decide whether adverse effects may arise and whether further testing is required.

Setting a general threshold for 'probably low risk' at 100 × EEC should not prevent an assessment of each pesticide with its particular characteristics. It must be recognized that not all harmful pesticides will be detected using this assessment scheme. Some effects may be too complex to be found with these test methods and trigger values.

Besides direct effects of pesticides on earthworms, we have to examine effects on predators, for which earthworms are an important food source. For example, cases where pesticide uptake by earthworms may cause secondary poisoning to birds have to be considered, involving assessment of possible uptake and accumulation of a pesticide in earthworms (Cooke *et al.*, in press). At present, ideas about how and when to assess this are at an early stage. Van Gestel & Ma (1988) pointed out that accumulation in earthworms might be very similar to accumulation in aquatic organisms so that special tests on earthworms would not give additional information. Further research is needed on this subject.

Besides the different kinds of toxic effects it must not be forgotten that some pesticides, especially herbicides, may have indirect effects on earthworms. Herbicides kill weeds and thus reduce food resources for earthworms during a long period of the year. Such effects are not considered in this assessment scheme but may be a very important factor for earthworm abundance in the field.

Risk management

The outcome of the testing procedure may be that a pesticide definitely has adverse effects on earthworms that are confirmed in a field test. The question then is what action has to be taken by the regulatory authority.

First, options for risk management to reduce the hazards has to be done. In the case of earthworms the possibilities are limited. Most pesticides are

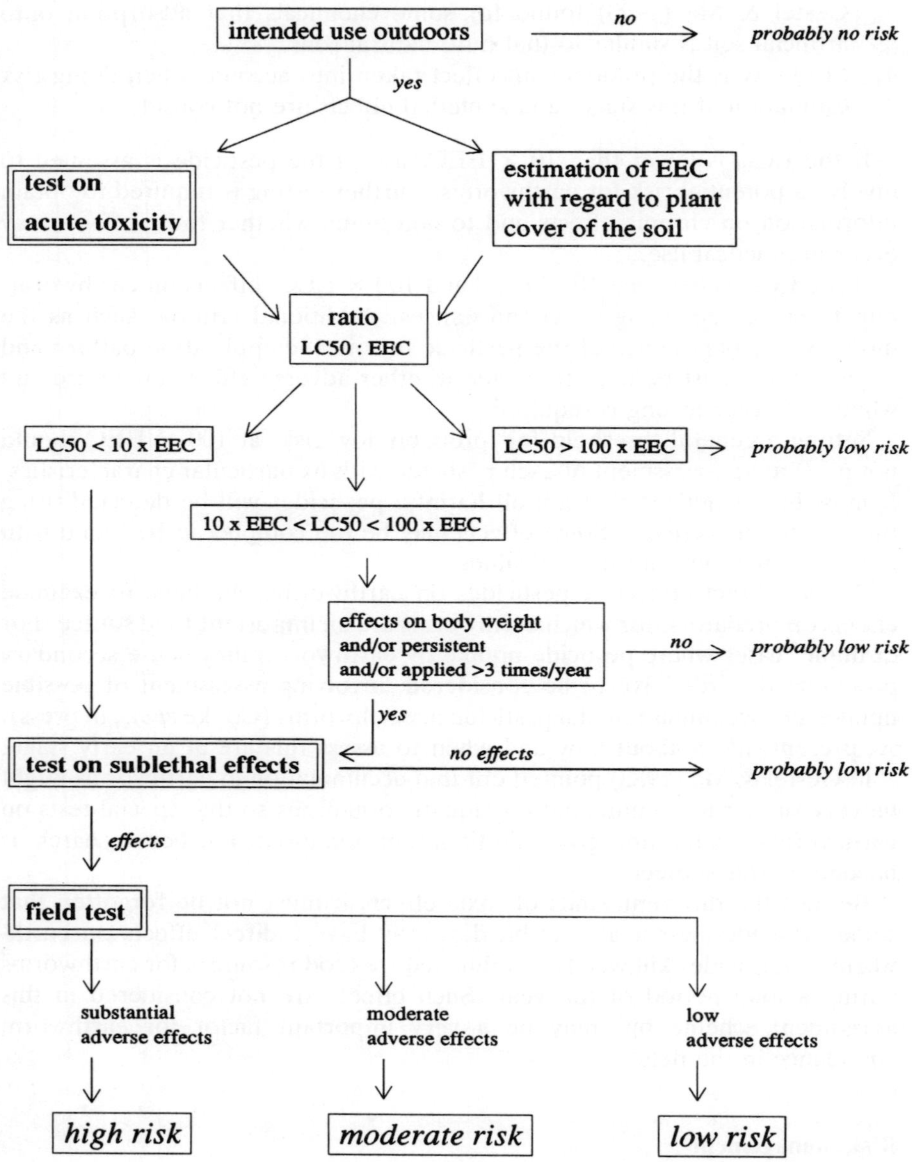

Figure 1. Proposal for a risk assessment scheme concerning effects of pesticides on earthworms

intended to be used where earthworms occur and therefore contamination of earthworms and their habitat can hardly be avoided.

After considering the possibilities for risk management, an expert decision is made on whether the effects are acceptable or not. So far there are no indisputable principles for such a decision. One criterion is to require that an effect is reversible within a reasonable time (Rothert *et al.*, 1990). When assessing the time for population recovery it has to be considered that earthworms have a long generation time and that immigration from unaffected areas may take place slowly.

One important reason for protecting earthworms is their contribution to soil fertility. Therefore it should be required at least that the pre-treatment population level is restored before the next application of the same pesticide.

One way of minimising the adverse effects is by a simple labelling, saying that the pesticide is harmful to earthworm, or giving information or restrictions that will reduce harm to earthworms. For example, it may be required that a pesticide which is harmful to earthworms should be applied only once a year or every two years. The amount of information that can be given on the label should be carefully considered. For example, it may be that especially earthworms of one ecological group suffer from the use of a pesticide. This is perhaps too much information for the label, but it might be helpful for an advisor from the local Plant Protection Service.

Outlook

There are still many questions concerning hazard assessment of pesticides on earthworms. For example the triggers for classifying a pesticide as 'persistent' until now are fixed from the chemist's point of view. What are relevant trigger values concerning earthworms that should be introduced into the assessment procedure?

The quantities of pesticide used in successive applications are added up without knowing what differences may exist between single and repeated doses. The significance of repeated small doses of pesticides for population dynamics needs further research.

Because of great variability in results the interpretation of data from field studies is very difficult. In addition, pesticides are often used together with others so that recovery time might be different in practical use.

Bioaccumulation of pesticides is not considered in routine testing. It should be introduced into assessment schemes for predators of earthworms.

It might be helpful to classify pesticides according to their overall adverse effects into the categories 'low risk', 'moderate risk' and 'high risk'. Important characteristics for classification should be the size and the duration of the effects.

References

BODE, E., BRASSE, D., KOKTA, C. (1988). Auswirkungen von Pflanzenschutzmitteln auf Nutzorganismen und Bodenfauna: Überlegungen im Zulassungsverfahren. *Gesunde Pflanzen,* **40**, 239–244.

BOUCHÉ, M.B. (1977). Strategies Lombriciennes. *Ecological Bulletin* (Stockholm), **25**, 122–132.

COOKE, A.S., GREIG-SMITH, P.W., JONES, S.A. (in press). Consequences for vertebrate wildlife of toxic residues in their earthworm prey. In: *Ecotoxicology of Earthworms* H. Becker, P.J. Edwards, P.W. Greig-Smith, F. Heimbach (eds) Intercept Press, Andover, Hants.

DUNGER, W. (1983). Tiere im Boden. *Die neue Brehm-Bücherei* **327**.

EDWARDS, P.J., BROWN, S.M. (1982). Use of grassland plots to study the effect of pesticides on earthworms. *Pedobiologia,* **24**, 145–150.

EDWARDS, C.A., LOFTY, J.R. (1977). Biology of earthworms. Chapman and Hall, London.

EEC (1985) EEC Directive 79/831. Annex V. Part C: Methods for the determination of ecotoxicity. Level 1. Toxicity for earthworms. Artificial soil test. DG XI/128/82.

HAQUE, A., EBING, W. (1983). Toxicity determination of pesticides to earthworms in the soil substrate. *Zeitschrift für Pflanzenkrankheiten und Pflanzenschutz,* **90**, 395–408.

HEIMBACH, F. (1985). Comparison of laboratory methods, using *Eisenia foetida* and *Lumbricus terrestris* for the assessment of hazard of chemicals to earthworms. *Zeitschrift für Pflanzenkrankheiten und Pflanzenschutz,* **92**, 186–193.

LOFS-HOLMIN, A. (1982). Reproduction and growth of common arable land and pasture species of earthworms (Lumbricidae) in laboratory cultures. *Swedish Journal of Agricultural Research,* **13**, 31–37.

MARTIN, N.A. (1986) Toxicity of pesticides to Allolobophora caliginosa. *New Zealand Journal of Agricultural Research,* **29**, 699–706.

OECD (1984) Guideline for testing of chemicals, No 207. Earthworms, acute toxicity tests. Adopted 4 April 1984.

ROTHERT, H., BRASSE, D., BODE, E. (1990). Abwägungs- und Entscheidungsprozesse bei der Zulassung von Pflanzenschutzmitteln im Hinblick auf deren Auswirkungen auf die terrestrische Fauna. *Gesunde Pflanzen,* **42**, 29–33.

SHIRES, S.W. (1985) A comparison of the effects of cypermethrin, parathion-methyl and DDT on cereal aphids, predatory beetles, earthworms and litter decomposition in spring wheat. *Crop Protection,* **4**, 177–193.

STRINGER, A., LYONS, C.H. (1974). The effect of benomyl and thiophanate-methyl on earthworm populations in apple orchards. *Pesticide Science,* **5**, 189–196.

VAN GESTEL, C.A.M., MA, W. (1988). Toxicity and bioaccumulation of chlorophenols in earthworms, in relation to bioavailability in soil. *Ecotoxicology and Environmental Safety,* **15**, 289–297.

VAN GESTEL, C.A.M., VAN DIS, W.A., VAN BREEMEN, E.M., SPARENBURG, P.M. (1989). Development of a standardized reproduction toxicity test with the earthworm species *Eisenia fetida andrei* using copper, pentachlorphenol, and 2,4-dichloraniline. *Ecotoxicology and Environmental Safety,* **18**, 305–312.

VAN GESTEL, C.A.M., MA, W. (1990). An approach to quantitative structure-activity relationships (QSARs) in earthworm toxicity studies. *Chemosphere,* **21**, 1023–1033.

Terrestrial Ecological Risk Assessment of Pesticides in the United States

A.M. STAVOLA AND H. CRAVEN

US Environmental Protection Agency, Washington DC, USA

Abstract

The U.S. Environmental Protection Agency (EPA) regulates pesticides by authority of the Federal Insecticide, Fungicide and Rodenticide Act (FIFRA). This process requires the ability to assess the risk of pesticides to determine if their use will cause 'unreasonable adverse effects on the environment'. Risk assessments of ecological effects are a necessary part of the process used to determine if a pesticide can be registered for a particular use. The Office of Pesticide Programs (OPP) developed a standard procedure describing the methods for conducting environmental risk assessments.

The methods used by OPP for conducting a terrestrial ecological risk assessment are discussed in this paper. It includes the ecological data requirements established by FIFRA, the quotient method of comparing hazard and exposure, weight-of-evidence review, including the significance of residues in earthworms and endpoints for regulatory decisions.

Introduction

The U.S. Environmental Protection Agency (EPA) has the authority to regulate pesticides through the Federal Insecticide, Fungicide and Rodenticide Act (FIFRA), as amended (7 U.S.C. 136 *et seq*). The Office of Pesticide Programs (OPP) has the responsibility for pesticide regulation. If it has been determined that the use of a pesticide will not cause unreasonable adverse effects on the environment then EPA can register that pesticide for specified uses. According to FIFRA the environmental, economic and social costs and benefits of the pesticide must be considered in making this determination. In order to achieve this decision a risk : benefit analysis must be performed. The focus of this paper is the process used by the Ecological Effects Branch (EEB) of the OPP to assess risk to terrestrial organisms.

Terrestrial ecological effects: testing requirements

Environmental risk assessment cannot be done until the ecological effects data required under FIFRA are submitted by the chemical companies applying for registration.

EPA has developed a tier system of ecological testing requirements. This approach allows data submission to progress from basic acute toxicity laboratory studies to chronic toxicity laboratory studies and then to complex field studies. In this approach the more complex studies are performed less often than the basic studies, because the need to conduct the higher tier studies is determined by the results of the lower tier studies.

AVIAN TESTING

The first tier for avian data requirements consists of short-term acute and subchronic toxicity studies in which the end point is mortality. These tests comprise one avian acute oral study and two avian subacute dietary studies conducted with the mallard duck (*Anas platyrhynchos*) and the bobwhite quail (*Colinus virginianus*). The test results are reported as LD_{50} values and LC_{50} values, respectively. One index may have a greater utility than the other, depending on the formulation or use pattern of the pesticide. For example, the LD_{50} is essential for risk assessments of seed treatments, baits and granular products. In contrast, LC_{50} values are critical when nongranular products are dispersed directly on avian food items.

The second tier consists of avian reproduction tests on mallard duck and bobwhite quail. These studies are required when the basic studies, environmental fate data and potential use patterns indicate that chronic effects may occur. The endpoints of these tests are mortality, growth and impairment of reproduction. The results of the chronic studies are reported as the NOEC (No Observed Effect Concentration) and the MATC (Maximum Allowable Toxicant Concentration). The NOEC indicates the test concentration at which there were no effects, and the MATC, which is often expressed as a geometric mean, describes the range between the NOEC and the next higher concentration at which effects were observed to occur.

The third and fourth tiers include simulated (pen studies) and actual terrestrial field testing, and they are designed to confirm or rebut the effects observed in any lower tier study by exposing birds to a pesticide under simulated or actual conditions. OPP prefers full field studies over pen studies as the former examine effects due to exposure by all routes to numerous species while eliminating artificial conditions imposed by the design of the pen studies. Field studies are necessary to negate EPA's assumption that the effects observed in single species laboratory studies are likely to occur under actual use conditions.

WILD MAMMAL TESTING

Testing to study the effects of pesticides on wild mammals is required on a case-by-case basis. For most pesticides, the rat acute oral LD_{50} study and the mouse reproduction study submitted for human toxicology risk assessment are sufficient to judge potential hazards to wild mammals. However, acute oral or subacute dietary studies on specific classes of mammals may be required if the other data are not sufficient to determine the potential hazard.

INSECT TESTING

Nontarget insect testing is required only if the proposed use will result in exposure of honey bees. The honey bee testing is based on the tier system – from acute contact LD_{50}, to tests of the toxicity to honey bees of residues on foliage, and field testing with pollinators. The field test has not yet been required.

PLANT TESTING

Testing of nontarget terrestrial plants is required for all herbicides and insecticides that are proposed for use in forests and natural grasslands. Testing of aquatic plants is required if the pesticide is to be applied aerially, or proposed for forestry sites or aquatic use.

Current terrestrial plant testing requirements consist of tests for seed germination/seedling emergence and vegetative vigor. The end points of these tests are changes in germination, growth and morphology, and the test results are reported as EC_{25} values (test concentrations at which 25 percent of the test population shows a reduction or change in the test parameter as compared to the control).

EARTHWORM TESTING

Toxicity tests on earthworms are not required. However, as discussed later, earthworms can be a component of the risk assessment.

Ecological risk assessment

OPP defines ecological risk assessment as the evaluation of the likelihood that the use of a pesticide will cause adverse effects such as mortality or reductions in populations of nontarget organisms due to acute, chronic or reproductive effects or disruption of community or ecosystem function (Urban and Cook, 1986). Ecological risk is a function of environmental exposure and toxicity.

QUOTIENT METHOD

In order to determine ecological risk, toxicological hazard and exposure need to be compared. This is known as the Quotient Method in which the Estimated Environmental Concentration (EEC) is compared to the toxicity level of concern. The resulting quotient is compared to regulatory risk criteria established by the Agency in 1975 in order to enforce FIFRA. Table 1 is

Table 1. Regulatory Risk Criteria for Terrestrial Vertebrates for Classification and Registrability of Outdoor Pesticide Applications

Non-Endangered			Endangered
Unrestricted use	Restricted use	Special review	
I. *Acute Toxicity*			
Nongranular Product:			
EEC <1/5 LC_{50} (Risk <0.2)	1/5 LC_{50} <EEC <LC (0.2 <Risk < 1)	EEC ≥LC_{50} (Risk ≥ 1)	EEC ≥ 1/10 LC_{50} (Risk ≥ 0.1)
Granular Bait/seed Product:			
[1]EEC <1/5 [2]LD_{50} (Risk <0.2)	1/5 LD_{50}<[1]EEC >[2]LD_{50} (0.2 <Risk < 1)	[1]EEC ≥ [2]LD_{50} (Risk ≥ 1)	[1]EEC ≥ 1/10 [2]LD_{50} (Risk ≥ 0.1)
II. *Chronic Toxicity*			
All Products:			
Chronic Exposure causes minor or no discernible effects	Chronic Exposure causes potentially significant effects but exposure is limited by restrictions	Chronic Exposure causes potentially significant reproductive impairment	Chronic Exposure causes potentially significant reproductive impairment
EEC < MATC (Risk < 1)	EEC ≥ MATC (Risk ≥ 1)	EEC ≥ MATC (Risk ≥ 1)	EEC ≥ MATC (Risk ≥ 1)

(Based on Urban and Cook, 1986)
[1] No. of granules/ft^2
[2] No. of granules equivalent to an LD_{50}

$$\text{Number of granules equivalent to an } LD_{50} = \frac{LD_{50}\ (\text{mg ai/kg} \times \text{Weight of animal (kg)})}{\%\ \text{ai of product} \times \text{Ave. weight of each granule (mg.)}}$$

based on the regulatory risk criteria in the Standard Evaluation Procedure (SEP) for Ecological Risk Assessment. Examination of the table will show that endangered species are afforded a wider margin of safety than are non-endangered birds and mammals. Endangered species are protected by federal law [The Endangered Species Act of 1973, as amended, (16 U.S.C. 1531, *et seq.*)], and even a single death of a species is considered highly significant.

The quotient method is a relatively simple approach to assess risk and clearly differentiates between chemicals that present a low risk to nontarget organisms and those that present a high risk. OPP acknowledges that there are several weaknesses with this method. It does not adequately account for effects of incremental dosages or indirect effects due to such things as food chain interactions. It does not adequately consider other effects in the ecosystem such as predator-prey relationships or structural shifts. Finally, it does not compensate for the differences between laboratory studies and the field populations, and it cannot quantify uncertainties. Therefore, this method cannot completely characterize the magnitude of the risk (Urban and Cook, 1986).

If the quotient equals or exceeds the Restricted Use criteria, an ecological risk is presumed. According to FIFRA this risk can be mitigated by limiting use of the pesticide to pesticide applicators who have been certified through EPA-approved training programs. In addition, the pesticide product is labelled to warn users of its toxicity to wildlife and to limit its entry into the habitats of endangered species.

However, if the quotient equals or exceeds the Special Review criteria (presumption of unacceptable risk) then a high ecological risk is presumed. In order to confirm or negate OPP's assessment of high risk, data confirming or negating the estimate of high risk are used. These data consist of terrestrial field studies.

Following is an example of how the quotient method was used to assess risk to birds exposed to highly toxic insecticides. Comparisons of the toxicity data with the exposure data for a granular insecticide, carbofuran, indicated it was likely to kill large numbers of birds when used according to proposed label directions as the estimated environmental residues greatly exceeded its LD_{50} values.

Birds can be expected to utilize pesticide-treated agricultural crops and, as a result, are likely to ingest pesticide granules. Birds will forage for seeds, grit, insects and soil invertebrates such as earthworms, particularly during the incorporation of granules into the soil. Often birds follow the planting equipment to feed on contaminated soil organisms turned up by the operation. While they are foraging they may ingest invertebrates containing high residue concentrations of the pesticide, or may ingest the granules themselves.

One fact unique to the risk assessment of carbofuran is that analysis of pesticide residues in earthworms was an important component of the dietary exposure evaluation. In a field study of granular carbofuran applied to corn, Balcomb *et al.* (1984) (as cited in Bascietto, 1987) reported whole-body carbofuran residues ranging from 0.3 ppm to 670 ppm, with a mean value of 85 ppm, in eleven of twelve samples of worms collected from treated furrows. In order to compare the residues found in the robin, *Turdus migratorius*, weighing 77 grams (Dunning, 1984 as cited in Bascietto, 1987) eats approximately ten percent of its body weight (7.7 grams) daily on a dry weight basis (Based on Kenaga, 1973 as cited in Bascietto, 1987). In spring about 80% of a robin's daily food intake (6.2 grams) is animal matter (Martin, *et al.*, 1951 as cited in Bascietto, 1987). The wet weight of an adult earthworm was estimated to be three to five grams, or 0.51 to 0.85 grams dry weight. Therefore, approximately seven 5-gram earthworms could be consumed daily by a robin requiring 6 grams dry weight animal matter. Based upon a mean carbofuran residue concentratiion of 85 ppm in earthworms, the dose in the seven worms is approximately 3 mg carbofuran.[1] This dose would be acutely toxic to robins as it is equivalent to a dose of 39 mg/kg[2] (Bascietto, 1987), which is 93 times greater than the LD_{50} value of 0.42 mg/kg for songbirds, (red-winged blackbird, *Agelaius phoeniceus*, Smith, 1987). Therefore it was concluded that deaths of songbirds are likely to occur if ingestion of carbofuran-contaminated earthworms constitutes a significant portion of their daily diet. Field studies were required to negate the assumption of high risk.

[1] 0.035 kg (wet weight of seven 5-gram worms) × 85 ppm (mg toxicant/kg worm) = 2.975 mg toxicant.

[2] 3 mg toxicant / 0.77 kg bird weight = 39 mg/kg

FIELD TESTING AND INCIDENT REPORTS

Accordingly, the highest tiers of ecological testing are not required until the quotient method predicts a high or unacceptable risk. Field studies can document whether or not the adverse effects predicted from laboratory data occur in the field from use according to label directions. These effects are not limited to mortality of birds or mammals, but may include effects on growth, reproduction and predator-prey relationships. The field studies are a method of attempting to address questions of risk that the quotient method is unable to address and to quantify the risk. However, due to the nature of the variable circumstances under which field studies can be conducted, the magnitude, frequency and duration of adverse effects such as mortality is probably greater than indicated by the numbers of dead carcasses found in a field study (Fite *et al.*, 1988).

Incident reports of mortality due to the use of pesticides are useful as confirmatory data. However, there is a great deal of under-reporting of field incidents due to a number of factors such as failure to observe the incident or failure to associate dead animals in a field with pesticide use. Therefore, the lack of field reports cannot be used as evidence that field effects do not occur.

In the case of carbofuran, field studies and a substantial number of incident reports were used as corroborative evidence to support our original prediction of high risk to birds. In one case, an intoxicated red-shouldered hawk (*Buteo lineatus*) which had been feeding on robins and voles in a field treated with Furadan 10G was discovered (Balcomb, 1983). This observation illustrated the importance of field data in depicting effects such as secondary poisoning that are not predicted by the quotient method of assessing risk.

WEIGHT OF EVIDENCE AND REGULATORY ENDPOINTS

All data are integrated in the final steps of the ecological risk assessment. The weight of evidence approach takes into account the sufficiency and quality of the data and the prevalence and magnitude of the observed or estimated effect. This entails weighing the quotient against conclusions derived from field studies and incident reports. Is the assumption of high risk derived from the quotient in which the EEC was compared to laboratory toxicity data still valid? Were the effects observed in the field marginal and transitory or were they substantial? Was there direct mortality or were the effects more subtle? In a terrestrial field study direct poisoning of birds and mamals may not be observed, but sublethal effects such as reduced fertility, nest abandonment, reduced food intake or loss of food sources may also be significant. The purpose of the weight of evidence review is to balance all these factors and reach a conclusion regarding the likelihood of the pesticide to cause unacceptable adverse effects when used in the environment. Any of these effects can be considered as regulatory endpoints if it is concluded that the effects are substantial.

These effects may be of a nature that cause reductions in local, regional or national populations or impact the stability of the community or ecosystem under consideration. Alternatively, they may be limited to effects that repeatedly and frequently kill nontarget organisms which may or may not lead to long-term population effects (Fite, *et al*, 1988). In either case, if the weight of evidence leads to the conclusion that the use of the pesticide will cause unacceptable adverse acute or chronic effects to nontarget organisms then the Agency can initiate a Special Review if the chemical is already registered for use as a pesticide. The outcome of the Special Review process is to reduce risk either through more label restrictions or by taking further regulatory action such as suspension or cancellation of the registration of the pesticide. If the chemical is not currently registered then the regulatory decision could be to deny registration.

Acknowledgements

The authors thank James Akerman and Douglas Urban for their comments and recommendations.

References

BALCOMB, R. (1983). Secondary poisoning of red-shouldered hawks with carbofuran. *Journal of Wildlife Management*, **47**, 1129–1132.

BALCOMB, R., BOWEN, C.A. III, WRIGHT, D., LAW, M. (1984). Effects on wildlife of at-planting corn applications of granular carbofuran. *Journal of Wildlife Management*, **48**, 1353–1359.

BASCIETTO, J.J. (1987). *Granular Carbofuran: Assessment of Risk to Wildlife*. Prepared for the Special Review of Carbofuran. U.S. Environmental Protection Agency, Washington, D.C., 164 pp.

DUNNING, J. (1984). Body weights of 686 species of North American birds. *Western Bird Banding Association, Monograph No. 1*.

FITE, E.C., TURNER, L.W., COOK, N.J., STUNKARD, C. (1988). *Guidance Document for Conducting Terrestrial Field Studies*. EPA 540/09–88–109. U.S. Environmental Protection Agency, Washington, D.C., 67 pp.

KENAGA, E.E. (1973). Factors to be Considered in the Evaluation of the Toxicity of Pesticides to Birds in Their Environment. In: *Global Aspects of Chemistry, Toxicology and Technology as Applied to the Environment. Environmental Quality and Safety*., F. Coulston and F. Korte (eds.), *vol. II*, Academic Press, NY, pp. 166–181.

MARTIN, A.C., ZIM, H.S., NELSON, A.L. (1951). *American Wildlife and Plants: A Guide to Wildlife Food Habits*. Dover Publications, NY, 500 pp.

SMITH, G.J. (1987). *Pesticide Use and Toxicology in Relation to Wildlife: Organophosphorus and Carbamate Compounds. Resource Publication 170*, U.S. Department of Interior, Fish and Wildlife Service, Washington, D.C. pp. 26–28.

URBAN, D.J., COOK, N.J. (1986). *Hazard Evaluation Division Standard Evaluation Procedure: Risk Assessment*. EPA–540/9–85–011. U.S. Environmental Protection Agency, Washington, D.C., 96 pp.

U.S. ENVIRONMENTAL PROTECTION AGENCY. (1982a). *Pesticide Assessment Guidelines Subdivision E: Hazard Evaluation: Wildlife and Aquatic Organisms*. EPA–540/9–82–024. U.S. Environmental Protection Agency, Washington, D.C., 86 pp.

U.S. ENVIRONMENTAL PROTECTION AGENCY. (1982b). *Pesticide Assessment*

Guidelines Subdivision J: Hazard Evaluation: Nontarget Plants. EPA–540/9–82–020. U.S. Environmental Protection Agency, Washington, D.C., 55 pp.

U.S. ENVIRONMENTAL PROTECTION AGENCY. (1982c). *Pesticide Assessment Guidelines Subdivision L: Hazard Evaluation: Nontarget Insects*. EPA–540/9–82–019. U.S. Environmental Protection Agency, Washington, D.C., 29 pp.

Chemical biomonitoring with earthworms

Assessment of Contaminated Soils using Earthworm Test Procedures

C.A. CALLAHAN[1] AND G. LINDER[2]

[1] *USEPA Environmental Research Laboratory, 200 S.W. 35th Street, Corvallis, OR 97333, USA*
[2] *ManTech Environmental Technology, Inc. USEPA Environmental Research Laboratory, 200 S.W. 35th Street, Corvallis, OR 97333, USA*

Abstract

Three approaches were used to assess the potential impact of soil contaminants at field sites such as Superfund Sites and dredge spoils site. Earthworms were used in laboratory bioassays, field bioassays, and as biological samples for measuring whole body residues for assessing impacts associated with various chemical contaminants. In relatively short, 7-day tests earthworms were adequate indicators of adverse impacts associated with organochlorine pesticides; for mixtures of PAHs, PCBs, and heavy metals, 28-day exposures appeared to be adequate to evaluate adverse impacts associated with these soil contaminants. Sublethal effects such as coiling, shortening, stiffening, swelling, and cutaneous lesions were indicators of earthworm morbidity. In conjunction with mortality data, these sublethal data provided quantitative measurements of biological effects associated with soils at contaminated field sites. Two earthworm species, *Eisenia fetida* Savigny and *Lumbricus terrestris* Linneaus yielded similar results in laboratory bioassays of soil samples. Laboratory bioassays also indicated that composite soil samples may underestimate soil toxicity compared to its constituent grab samples.

In evaluating tissue residues of target contaminants, differences in contaminant uptake were observed among the soil samples; however, field and laboratory uptake for metals and organics closely corresponded. Native species collected on site did not significantly differ with respect to chemical uptake from *Eisenia fetida* which was used for on-site testing. The period of time for on-site exposure is critical, however, if equilibrium conditions between earthworm tissue residues and soil contaminant levels are targeted objectives for on-site testing. Earthworms were also sufficiently sensitive to detect spatially variable soil contaminant concentrations throughout various habitats at the field site, and biological effects observed in laboratory tests with site-soil samples closely corresponded with the effects observed in the field.

Introduction

Soil contamination evaluations are frequently required in hazardous waste site assessments. Under the regulatory auspices of the Comprehensive Environmental Response Compensation and Liability Act of 1980 (CERCLA), as amended by the Superfund Amendment and Reauthorization Act of 1986 (SARA), and the Response Conservation and Recovery Act (RCRA) of 1976, these soil contamination studies have increasingly relied upon biological tests for developing site-specific human health and ecological risk assessments. To complicate the risk assessment process, however, these sites are frequently characterized by complex chemical mixtures rather than single chemicals, and exposure may be multi-media in character and involve inhalation, ingestion, and dermal routes. Certain groups of chemicals such as pesticides, metals, and organic solvents have repeatedly been identified as target contaminants at various sites. But, in evaluating adverse biological effects associated with the site, the isolation and identification of any single constituent chemical of the mixture as a causal agent is often difficult, if not impossible, because of the complex exposure characteristics and variably expressed biological response in target populations. These site characteristics have subsequently encouraged the use of a toxicity-based approach rather than a chemically-based approach to determine the potential and actual environmental impact associated with complex chemical mixtures at the site. While chemically-based approaches are solely dependent upon extensive chemical analysis of environmental samples, toxicity-based approaches directly integrate biological information into the assessment process (Parkhurst, *et al.* 1989). In a toxicity-based strategy, bioassays and toxicity tests directly measure contaminant bioavailability, as well as any associated adverse biological and toxic effects. With biological testing, affects such as reduced growth, reduced reproduction, morbidity, and death may be associated with exposure to complex mixtures in instances when the mechanism of toxicity may not be readily apparent, and the specific cause of the effect is unknown.

For ecological assessments at hazardous waste sites, soil macroinvertebrates are potential targets frequently evaluated as part of a soil contamination evaluation. Earthworm tests extend critical contributions to these soil evaluations, since the methods directly test the soil matrix. When laboratory toxicity assessments with earthworms are combined with chemical analyses of soil and biological samples, and field tests which measure *in situ* effects on earthworms, higher level biological organization (e.g., populations and communities) may be more adequately evaluated during the risk assessment process.

To demonstrate the assessment strategy for a hazardous waste site, a case history will be reviewed which illustrates the toxicity-based approach in soil contamination evaluations. The three components of the approach have been used in this case study, and include laboratory testing for acute effects, field or on-site testing for evaluating *in situ* effects, and measurement of the chemical toxicant in the soil matrix from the site. The data derived from these component parts contributes critical information that is subsequently integrated to the site assessment process.

Soil contamination evaluation: a case study

BAIRD & MCGUIRE SUPERFUND SITE, HOLBROOK, MA. SITE HISTORY

The Baird & McGuire Superfund Site is located approximately 20 miles south of Boston in Holbrook, MA and is characterized by several habitats that have been exposed to complex mixtures of persistent chemicals. For nearly 60 years, the Baird & McGuire Company used the site to mix and batch agrochemicals, including herbicides and pesticides for resale. Pesticides, herbicides, fungicides, and polycyclic aromatic hydrocarbons are known to have been spilled or dumped over almost the entire site during the operation of the company, and preliminary monitoring found both 'hot spots' and widespread chemical contamination at the site. As a general rule, the level of pesticides found on the site are much higher than would be found under conditions of agricultural applications (Korschgen 1970).

PRELIMINARY STUDY DESIGN AND METHODS: APPLICATION OF THE TOXICITY-BASED APPROACH

Detailed study designs, data compilations, and a complete site assessment are summarized in Callahan, *et al.* (1991). In brief, the history of the Baird & McGuire site indicated that the site could be stratified into sampling units characterized by habitat and land use patterns (Figure 1). Organochlorine pesticides were targeted as contaminants of concern and were the primary analytes for soil and tissue residue analysis. Because of the existing site information and the past work regarding pesticide impacts on earthworms (Drewes, *et al.* 1988; Haque and Ebing, 1983; Wheatley and Hardman, 1968), mortality and morbidity (sublethal endpoints such as coiling, shortening, stiffening, swelling, and cutaneous lesions) were identified as measurement endpoints for the biological evaluations. These data were collected during field tests using *Lumbricus terrestris* Linneaus, as well as during laboratory earthworm tests. Also, based upon previous land use information, soil samples were collected from sampling units which presumedly had 'high,' 'intermediate,' and 'low' concentrations of the targeted contaminants (FSSW-1, FSSW-2, and FSSW-3, respectively), as well as a reference site (see Figure 1). Samples collected from each of these sampling units were split for quantification of target analytes as well as for laboratory bioassays. In the laboratory, the 14-day test (Greene, *et al.* 1989) using *Eisenia fetida* Savigny was adapted to evaluate site-soil samples. Also, a 28-day laboratory test

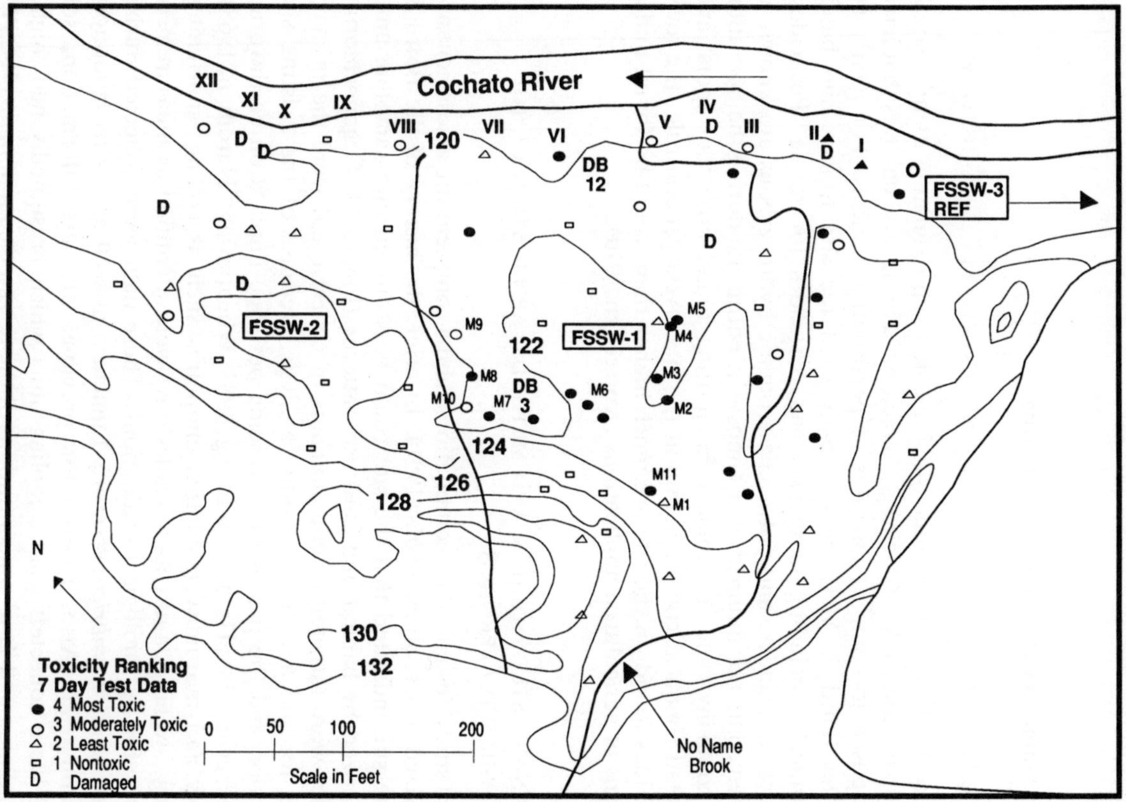

Figure 1. Stratified sampling design at Baird & McGuire Superfund site (FSSW–1, FSSW–2, and FSSW–3 plus reference location) with on-site transects, and the spatial distribution of biological responses summarized as rank scores (after Callahan *et al.* 1991).

(Food & Drug Administration [FDA], 1987) using *L. terrestris* was performed on soil-sample splits (Springborne Laboratories; Wareham, Massachusetts).

For the field test, containers were placed on transects located across the site in the impacted, as well as a reference areas. These plastic exposure chambers were placed into soil excavations located along the transects, then the exposure chambers were filled with the excavated soil, and five *L. terrestris* were placed on the soil surface.

Results and discussion

INTEGRATED LABORATORY AND FIELD STUDIES

Given the site-specific conditions briefly outlined above, three tasks had been identified as pertinent to the Baird & McGuire risk assessment process, and these were performed during 1988 as part of a soil contamination evaluation:

- laboratory toxicity testing with *Eisenia fetida* and *Lumbricus terrestris* was performed on soil samples collected at various locations at Baird-McGuire;
- on-site field tests were completed with *Lumbricus terrestris* at those same locations; and
- targeted chemical contaminants were measured in soil samples obtained from the field test locations, as well as from whole earthworms exposed in the on-site test.

During 1989, mortality and the morbidity endpoints, i.e., coiling, shortening, stiffening, swelling, and cutaneous lesions, were evaluated at the end of 7-day field exposures using *L. terrestris*. Soil samples as well as earthworm tissue samples from survivors exposed to those soil samples were randomly selected for chemical analysis. In total, 36 sampling locations had soil contaminant measurements taken, and 29 locations had sufficient survivorship for quantification of the targeted organochlorine analytes in earthworm tissues (Callahan, *et al.* 1991).

LABORATORY TESTS

Two bioassays were performed, the 14-day *Eisenia fetida* test (Callahan, *et al.* 1985) and the 28–day *Lumbricus terrestris* test which followed an FDA protocol (FDA, 1987). The 14-day *Eisenia* test results indicated that the greatest mortality consistently occurred in grab samples collected on the site. For example, composite samples from FSSW-1 and FSSW-2 expressed less toxicity than tests completed with their constituent grab samples. The reference and FSSW-3 soil samples expressed no significant acute toxicity. The results of the 28-day *Lumbricus* laboratory test reinforced the results of the *Eisenia* laboratory test, since the split samples displayed the same trend for mortality data across all samples collected on-site (Figure 2). Furthermore, these test results for organochlorine soil contaminants were consistent with previous comparisons regarding these same earthworm

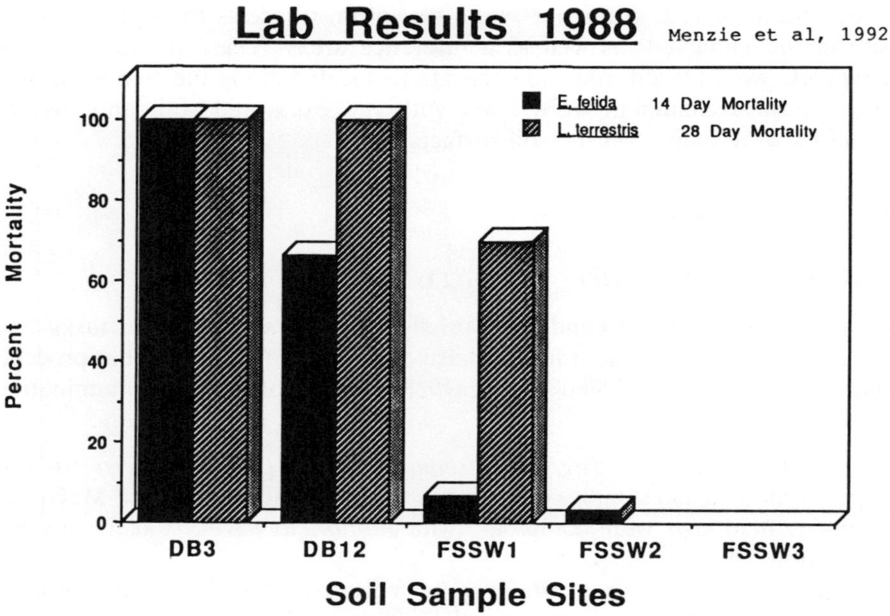

Figure 2. Mortality of *Eisenia fetida* (solid bars) and *Lumbricus terrestris* (hatched bars) exposed to split soil samples collected from sampling areas identified in Figure 1.

species when exposed to these chemicals in single-compound exposures (Heimbach, 1985).

Apparently, the choice of *Lumbricus terrestris* or *Eisenia fetida* makes little difference for evaluating the results of on-site test results with those of the laboratory. This appears to be true for heavy metals, PCBs, PAHs, and organochlorines, although each have their own characteristic uptake rates and maximum tissue concentrations. Both species can exist for relatively long periods in field situations as well as laboratory conditions. However, as might be expected, basic requirements of moisture and temperature have been shown to be necessary for survival. There can be great differences between native worms and those placed on-site with respect to the rates of uptake and the kinds of contaminants taken up.

FIELD TESTS

In the field test completed with *L. terrestris*, 100% mortality was recorded after 7-day exposures at some locations. However, marked spatial heterogeneity, which was not unexpected, was indicated, since there were some testing locations without observed morbidity. Some locations consistently indicated that the soil was acutely toxic, and clearly was associated with high mortality rates in exposed earthworms. For example, the on-site test completed with *L. terrestris* indicated that there were ‘hot spots’ in highly contaminated areas (e.g., FSSW-1). Similarly, comparisons among earthworm body burdens of target analytes, soil concentrations of target analytes, and sublethal response data (morbidity endpoints such as coiling, shortening, stiffening, swelling, and cutaneous lesions) were made, and associations between biological information (sublethal responses and tissue residues) and soil contaminant sources were identified.

Soil concentrations of targeted analytes suggested that chlordane was consistently present at the lowest concentrations across the site, but DDT generally occurred at the highest concentrations. For example, DDT concentrations in some site-soil samples exceeded 1000 ug/g (Callahan, *et al.* 1991). Mortality was highly correlated to tissue and soil concentrations of chlordane, DDT and its metabolites (DDTR), whereas none of the morbidity endpoints were significantly correlated to these soil contaminants.

Some sublethal responses, however, were correlated to tissue residues of the targeted analytes. Swelling, stiffening, and coiling were significantly correlated to earthworm tissue DDTR but not to soil DDTR. Similarly, swelling was significantly correlated to earthworm chlordane residue concentrations but not to soil chlordane. Lesions were not correlated to either chlordane or DDTR tissue residues, nor were cutaneous lesions correlated with soil concentrations of these analytes. And, while tissue samples collected from earthworms exposed to reference soils presented no detected residues of targeted chemicals, detectable amounts of DDT, DDD, and DDE were measured in the corresponding soil sample.

The sublethal responses and mortality data collected from earthworms exposed in containers at Baird & McGuire were used to rank the sampling locations based on the biological responses associated with exposure to site-soil samples (Callahan *et al.* 1991). The spatial distribution of these biological responses over the site and the absence of similar responses in the reference area indicated that adverse biological effects were associated with the contaminants found in soils collected at Baird & McGuire (see Figure 1).

Summary and Conclusions

A primary reason for assessing contaminated sites is to establish whether an impact has occurred and to establish, if possible, whether that impact is due to contamination or some physical alteration of the habitat. The process for answering such questions involves measuring biological impacts through bioassays in the laboratory and the field, and a measurement of contaminant concentrations through chemical analyses of the same matrix (Parkhurst, *et al.* 1989; Charters, 1990). As representative soil macroinvertebrates, earthworms have been useful biological assessment 'tools' in many soil contamination evaluations, For example, in soil contamination evaluations, laboratory and field tests with earthworms can help establish linkages between contaminants and adverse effects observed in soil organisms.

Laboratory tests of site-soils have previously been completed by modifying a laboratory method (Edwards 1984) that exposed earthworms for 14 days to mixtures of site soils and artificial soil (Callahan, *et al.* 1985). In this modified method, acute toxicity is estimated as a median lethal concentration (LC_{50}) which represents the percentage of site soil to artificial soil (wt: wt) that produces mortality in 50% of the exposed animals. Although mortality has been the most frequently reported endpoint, other measures of impact have also been successfully used to demonstrate sublethal contaminant effects in earthworms. Measurement endpoints such as whole body concentration of chemicals (Simmers, *et al.* 1987) and suppressiion of the giant axon conduction velocity (Drewes, *et al.* 1988; Callahan, *et al.* 1985) have been shown to be effective measures of impact. Also, endpoints which measure immunosuppression in earthworms are being developed at the University of North Texas (see these proceedings). Regardless the measurement endpoint, when a rigorous sampling scheme is used to collect site-soils, the test results can be used to delineate patterns of toxicity at the site (Thomas, *et al.*, 1986; Callahan, *et al.* 1991).

Field assessments of soil contamination can be successfully completed through on-site testing with earthworms. In the Baird & McGuire study reviewed here, laboratory evaluations corresponded with field test results for mortality and uptake of contaminants. These on-site assessments, however, reduced the reliance upon extrapolations from laboratory results to field effects for the interpreting the contaminant impacts on test species. Field assessments often involve native species found on the site or surrogate earthworm species placed on site which reduces the uncertainty associated with laboratory-to-field extrapolations. Various test modifications may be implemented to address this uncertainty. The OECD 14-day laboratory method using *Eisenia fetida*, for example, may be modified for soil contamination evaluations. For Baird & McGuire soil samples, the adverse effects observed in test with *Eisenia fetida* compared favorably with those results noted for *Lumbricus terrestris*. However, as noted here, differences between these test methods may be apparent when soil sampling methods varied, e.g., composite versus grab samples. There is ample support, then, for both field and laboratory methods, depending upon the site-specific questions being addressed.

Overall, the lumbricid earthworm, *L. terrestris*, appears to be a sensitive organism for assessing the potential impact of contaminated organochlorine soil at field sites. While other field evaluations (Marquenie, *et al.* 1987; USEPA 1991) used *Eisenia fetida* for longer exposure periods (28 days and 14 days, respectively) to determine site-specific bioaccumulation potentials of heavy metals and polynuclear aromatic hydrocarbons (PAH), *L. terrestris* displayed significant correlations between soil contaminants and tissue residues following these 7-day exposures to organochlorine pesticides.

References

CALLAHAN, C.A., RUSSELL, L.K. AND PETERSON, S.A. (1985). A comparison of three earthworm bioassay procedures for the assessment of environmental samples containing hazardous wastes. Biol Fert Soils 1: 195–200.

CALLAHAN, C.A., MENZIE, C.A., BURMASTER, D.E., WILBORN, D.C. AND ERNST, T. (1991). On-Site Methods for Assessing Chemical Impact on the Soil

Environment using Earthworms: A Case Study at the Baird & McGuire Superfund Site, Holbrook, MA. Environ. Tox. and Chem. Vol 10(6), 817–826.

CHARTERS, D.W., (1990). The Use of Toxicity Tests in Environmental Assessments at Superfund Sites, Abstract no. 164. SETAC Presentation at Annual Meeting, Nov. 11–15, Arlington, VA.

DREWES, C.D., VINING, E.P. AND CALLAHAN, C.A. (1984). Non-invasive electrophysiological monitoring: A sensitive method for detecting sublethal neurotoxicity in earthworms. Env. Tox. and Chem. Vol 3. pp 599–607.

DREWES, C.D., VINING, E.P. AND CALLAHAN, C.A. (1988). Electrophysiological detection of sublethal neurotoxic effects in intact earthworms. In: C.A. Edwards and E.F. Neuhauser (Eds) Proceedings of the International Conference on Earthworms in Waste and Environmental Management, July, 1984, Cambridge, England, pp 355–366, SPB Academic Pub., The Haque, The Netherlands.

EDWARDS, C.A. (1984). Report of the second stage in development of a standardized laboratory method for assessing the toxicity of chemical substances to earthworms. Commission of the European Communities. EUR 9360 EN. 98 pp.

FDA (1987). Environmental assessment technical assistance document 4.12. Earthworm subacute toxicity. Food and Drug Administration, Center for Food Safety and Applied Nutrition Environmental Impact Section and Center for Veterinary Medicine Environmental Staff. Washington D.C.

GREENE, J.C., BARTELS, C.L., WARREN-HICKS, W.J., PARKHURST, B.R., LINDER, G.L., PETERSON, S.A. AND MILLER, W.E. (1989). Protocols for short term toxicity screening of hazardous waste sites. EPA/600/3–88/029, USEPA ERL-Corvallis, OR.

HAQUE, A. AND EBING, W. (1983). Toxicity determination of pesticides to earthworms in the soil substrate. Journal of Plant Diseases and Protection. 90(4): 395–408.

HEIMBACH, F. (1985). Comparison of laboratory methods, using *Eisenia foetida* and *Lumbricus terrestris*, for the assessment of the hazard of chemicals to earthworms. Jour. of Plant Dis. and Prot. 92(2): 186–193.

USEPA. (1991). Final Technical Report for the Preliminary Field Survey and On-Site, *In Situ* and Laboratory Evaluations Completed at Milltown Reservoir [FY 90]. USEPA, Environmental Research Laboratory, Corvallis, OR.

KORSCHGEN, L.J. (1970). Soil-food-chain-pesticide wildlife relationships in aldrin-treated fields. Jour. Wildl. Manage. Vol 34(1) 186–199.

MA, W., EDLEMAN, TH., VAN BEERSUM, I. AND JANS, TH. (1983). Uptake of cadmium, zinc, lead, and copper by earthworms near a zinc-smeltering complex: Influence of soil pH and organic matter. Bull. Environ. Contam. toxicol. 30: 424–427.

MARQUENIE, J.M., SIMMERS, J.W. AND KAY, S.H. (1987). Preliminary assessment of bioaccumulation of metals and organic contaminants at the Times Beach confined disposal site, Buffalo, N.Y. Misc. Paper EL–87–6. Waterways Experiment Station, Corps of Engineers Vicksburg, MS.39180–0631.

MENZIE, C.A., BURMASTER, D.E., FRESHMAN, J.S. AND CALLAHAN, C.A. (1992). Assessment of Methods for Estimating Ecological Risk in the Terrestrial Component: A Case Study at the Baird & McGuire Superfund Site in Holbrook, Massachusetts. Environ. Toxicol. and Chem. Vol. 11(2), 245–260.

PARKHURST, B.J., LINDER, G., MCBEE, K., BITTON, G., DUTKA, B. AND HENDRICKS, C.W. (1989). Toxicity Tests, Chap. 6. In: Ecological Assessment of Hazardous Waste Sites: A Field and Laboratory Reference. Eds: W. Warren-Hicks, B.R. Parkhurst, and S.S. Baker EPA/600/3–89/013, ERL-Corvallis.

PUBLIC LAW 94–590. (1976). Resource Conservation and Recovery Act (RCRA), as amended.

PUBLIC LAW 96–510. (1980). Comprehensive Environmental Response, Compensation, and Liability Act (CERCLA), as amended.

PUBLIC LAW 94–499. (1986). Superfund Amendment and Reauthorization Act (SARA), as amended.

THOMAS, J.M., SKALSKI, J.R., CLINE, J.F., MCSHANE, M.C., SIMPSON, J.C., MILLER, W.E., MILLER, S.A., CALLAHAN, C.A. AND GREENE, J.C. (1986). Characterization of Chemical Waste Site Contamination and Determination of its Extent using Bioassays. Environ. Toxicol. and Chem. Vol 5, 1–15.

WARREN-HICKS, W., PARKHURST, B.R. AND BAKER, S.S. (1989). Ecological Assessment of Hazardous Waste Sites: A Field and Laboratory Reference. Introduction, Chapter 1. EPA/600/3–89/013 ERL-Corvallis.

WHEATLEY, G.A. AND HARDMAN, J.A. (1968). Organochlorine insecticide residues in earthworms from arable soils. J. Sci. Fd. Agric. Vol 19(4) 221–225.

Earthworms as Indicators of Ecotoxicity

B.J. VENABLES[1,2], L.C. FITZPATRICK[1] AND A.J. GOVEN[1]

[1] *Environmental Effects Research Group, Division of Environmental Sciences, Department of Biological Sciences, University of North Texas, Denton, TX 76203, USA*
[2] *TRAC Laboratories, Inc., 113 Cedar St., Denton, TX 76201, USA*

Abstract

Development and use of a comprehensive panel of phylogenetically-conserved biomarkers in earthworms are discussed in terms of criteria for good measurement endpoints, tier approach for immunotoxicology and application to toxicity assessment of hazardous waste-site soils.

Introduction

As other papers within this workshop attest, earthworms have been used extensively in toxicity studies, primarily those involving agrochemicals. Although the importance of using earthworms in environmental toxicology seems obvious, their full potential has not been recognized. Most work still emphasizes establishing LC50s for chemicals associated with soil contamination. Development of sublethal assays, until recently, has been largely ignored. We believe that, as a single taxon, earthworms have great potential as non-mammalian surrogates for use in: (1) assessing sublethal risks to public and environmental health from contaminated soils and hazardous waste sites (HWS); (2) preliminary screening of neat chemicals; and (3) the evaluation of the toxicity of complex mixtures. Thus, we have been developing a comprehensive panel of sublethal measurement endpoint biomarkers, based on sensitive phylogenetically-conserved processes, within the framework of the National Toxicology Program's (NTP) two-tiered immunotoxicity protocol for mice (Luster *et al.*, 1988) and according to established criteria for good measurement endpoints (Suter, 1989).

We have been evaluating the biomarkers in *Lumbricus terrestris* and *Eisenia foetida* using PCBs, pesticides, heavy metals, and complex mixtures in HWS soils and combustion residues (Goven *et al.*, 1988; Venables *et al.*, 1988; Rodriguez *et al.*, 1989; Fitzpatrick *et al.*, 1992). Our goal is to document that whole-earthworm respones, using an integrated suite of key physiological and biochemical assays, can provide sensitive predictive indices of both toxicity mechanisms and potential risks to higher animals, including humans. Our approach is to determine comparative tissue-specific dose/response profiles of the individual assays to classes of xenobiotics and pollutants (i.e. chemical structure-response relations), and compare them with those in mice (i.e. determine earthworm dose-equivalents necessary to produce analogous/homologous responses in mice).

Here we briefly discuss earthworm biomarkers, criteria for good measurement endpoints, the tier approach for immunoassays and an application of several assays to HWS soils using *L. terrestris*.

Characteristics of Good Indicators

Biological indicators of chemical stress should be sensitive and stable. They must show variability in response to different pollutants, but consistent response to non-pollutant environmental changes (e.g. habitat change) to maximize signal-to-noise ratio (i.e. pollutant response: natural variability). Additionally, they should measure something of ecological significance (or directly related to something of ecological significance), be easily and unambiguously quantifiable, biologically appropriate for expected exposure routes and, if possible, indicative of specific classes of pollutants. Table 1 contains a list of characteristics (modified from Suter, 1989) which have guided our selection of specific assays and measurement endpoints.

Table 1. Criteria for good measurement endpoints

1.	Correspondence to or predictive of assessment endpoints.
2.	Broadly applicable.
3.	Readily measured and standardized.
4.	Appropriate to HWS scale, exposure pathway and temporal dynamics.
5.	Low natural variability.
6.	Diagnostic of relevant pollutants.

CANDIDATE ENDPOINTS FOR EARTHWORMS

Characteristics of good measurement endpoints depend on the level of biological organization being examined. Responses at the population level are ecologically meaningful, but have high natural variability and are not sufficiently sensitive to be diagnostic of specific pollutants. Biochemical endpoints are sensitive and easily standardized, but are not easily used to extrapolate to meaningful ecological consequences. Therefore, to be comprehensive, a system should include indicators from a broad range of levels of biological organization. We are currently evaluating a variety of candidate indicator endpoints which range from the molecular to the population level (Table 2). Many of these candidate indicators are under development in our laboratories, while others have been developed elsewhere. Herein, we focus on sublethal assays, principally immunological, and their organization within the tier approach.

Table 2. Candidates for an integrated earthworm-based panel of ecological indicators.

Population	**Reproductive/Developmental**	**Immunological**
Demography	Sperm Counts	Non-Specific
Natality/Mortality	Sperm Viability	Phagocytosis
Density	Spermatogenesis	NBT Reduction
Whole Earthworm	Oogenesis	Lysozyme
LC/LD50	Cocoon Production	Super Oxide
Morbidity	Hatchability	Chemiluminesence
VO_2	Teratogenesis	N K Cells
Mass loss	Growth/Maturation	Wound Healing
Bioavailability/Biocentration	**Neurological**	Inflammation
Uptake/Depuration	Giant Axon Conductivity	Humoral Mediated
Molecular	**Coelomocyte**	Lytic/Agglutination Titers
DNA Adducts	Total Cell Counts	Cell Mediated
'Stress' Proteins	Differential Cell Counts	Mixed Leukocyte Response
Protein Synthesis	Cell Viability	Blast Transformation
	Repopulation Rate	Graft Rejection
	Steriology of Organelles	Host-Challenge

Immunoassays as indicators

Although many methods have shown toxic effects on the immune system, only relatively recently have scientists become aware of the broad spectrum of xenobiotics that alter immune function (e.g. Vos, 1977; Sharma, 1981; Dean *et al.*, 1985) and of its potential as a target organ system for studying chemical toxicity (Dean *et al.*, 1986). Since much is known about the molecular and cellular biology of the immune system, it is especially well-suited for studying mechanisms of toxicity. Cells of the immune system have a number of phylogenetically conserved characteristics which make them appropriate and sensitive for studying chemical-induced cellular toxicity: (1) capacity to proliferate rapidly following activation with antigen or non-specific stimuli; (2) gene products that can be used as markers of maturation; and (3) potential to undergo terminal differentiation that results in production of measurable humoral mediators (e.g. antibodies in mammals and antisomes in earthworms) or providing effector functions (Luster *et al.*, 1988). The immunobiology of earthworms has been studied intensively providing the basic foundation for developing assays for immunotoxicological studies (see Fitzpatrick *et al.*, 1990).

TIER APPROACH TO ASSESS IMMUNOTOXICITY

Complexity of the immune system and the fact that no single immunoassay is sufficient for screening chemicals or assessing mechanisms of toxic action, have led to tiered approaches using multiple assays and endpoints (Vos, 1980; Dean *et al.*, 1982; Dean *et al.*, 1986; Luster *et al.*, 1988). A typical tier approach involves two panels of immunoassays: (1) A panel of relatively simple parameters of cell-mediated immunity (CMI), humoral-mediated immunity (HMI), non-specific immunity (NSI) and immunopathology, which with limited effort can be used to screen xenobiotics; and (2) a comprehensive panel of additional CMI, HMI, NSI, immunopathology and host-resistance

Table 3. Earthworm tier I immunoassays for use in screening chemicals for immunomodulatory effects with corresponding murine immunoassays.[1]

PARAMETER	PROCEDURE Earthworm	Mouse
Immunopathology		
Cytology	Complete/differential coelomocyte count	Complete/differential leukocyte count
Body Mass	Total body weight	Total body weight
Histology	Coelomatopoietic tissue (Somatopleure)	Spleen, thymus, lymph node
Inflammatory	Wound healing	Wound healing
Humoral Mediated Immunity		
Agglutinins/Lytic Factors	Agglutination/lysis (erythrocytes)	Agglutination (erythrocytes)
Rosette formation	Secretory rosette (SR)	B-cell rosetteing
Plaque assay	Lytic factor erythrocyte PFCs[2]	IgM anti-erythrocyte PFCs[2]
Cell-Mediated Immunity		
Allo-antigen recognition	Allogeneic tissue recognition/ rejection	Allogeneic tissue recognition/rejection
Non-Specific Immunity		
Natural killer cell activity	Allogeneic/xenogeneic target cells	Allogeneic/xenogeneic target cells
Phagocytosis	Erythrocytes/fluorescent beads	Erythrocytes/fluorescent beads

[1] Modified from NTP Tier I Protocol (Luster *et al.*, 1988) According to Putative Earthworm Murine Correspondence.
[2] PFC = Plaque forming cells

challenge endpoints, that enable further testing of those chemicals producing effects in Tier I, usually at lower dose levels, and to assess modes of action and specific nature of toxic effects. According to Luster *et al.*, (1988) '. . . no compound has been found to affect an assay in Tier II without demonstrating some effect in Tier I.' The tier approach enables efficient cost-effective assessment xenobiotic immunotoxicity.

Our approach to develop immunoassays using earthworms follows the NTP's two-tiered immunoassay protocol using mice (Luster *et al.*, 1988). The NTP's objective was to develop a panel of immunoassays having sufficient sensitivity and comprehensiveness to detect subtle immunologic effects from exposure to chemicals, and relate them to alterations in host resistance to disease-producing organisms (e.g. bacteria, viruses, parasites) or tumor cells. Since diagnosis or prediction of secondary immunodeficiency effects (i.e. disease) following initial immunologic lesions from chemical exposure would be valuable in public health risk assessment, the NTP emphasized sensitive infectivity models and correlation between immunologic effects and altered host resistance. Additionally, miniaturization, simplicity and reproducibility were criteria used in selection of immunossays for the NTP protocol. According to Luster *et al.*, (1988), the NTP protocol has been effective in detecting clear in vivo dose-responses for prototype chemicals, and shows promise in correlating functional immunologic measurements with host-resistance challenge models.

In our work, coelomocytes from earthworms exposed to chemicals initially are assayed using Tier I tests (Table 3). While the probability of detecting immunotoxicity in Tier I is high, only limited information on the specificity of the immune defect may be discernable. For this reason, chemicals causing measurable effects in Tier I are re-examined using only relevant,

Table 4. Earthworm tier II immunoassays for comprehensive analysis of immunomodulatory effects of chemicals/drugs with corresponding murine immunoassays.[1]

PARAMETER	PROCEDURE Earthworm	Mouse
Immunopathology		
Immune cell quantitation	Flow cytometric analysis of coelomocytes	Flow cytometric analyses of circulating/peritoneal leukocytes
Humoral Mediated Immunity		
Challenge response	Erythrocyte agglutination	Erythrocyte agglutination
	Bacterial agglutination	Bacterial agglutination
	Lytic factor erythrocyte PFCs[2]	IgG anti-erythrocyte PFCs[2]
Antimicrobial action	Bacteriostatic effect of coelomic fluid	Bacteriostatic effect of serum
Cell-Mediated Immunity		
Blast-transformation	Mitogen/antigen blastogenesis	Mitogen/antigen blastogenesis
Mixed leukocyte response	Allogeneic/xenogeneic stimulation	Allogeneic/xenogeneic stimulation
Non-Specific Immunity		
Phagocytosis	Coelomocyte bacterial ingestion	Peritoneal cell bacterial ingestion
Bacterial activity	Coelomocyte bacterial killing	Peritoneal cell bacterial killing
Enzyme function	NBT[3] dye reduction by coelomocytes	NBT[3] dye reduction by peritoneal cells
Host Resistance Challenge Models		
Bacterial Models	Resistance to bacterial challenge	Resistance to bacterial challenge

[1] Modified from NTP Tier I Protocol (Luster *et al.*, 1988) According to Putative earthworm Murine Correspondence.
[2] PFC = Plaque forming cells
[3] NBT = Nitroblue tetrazolium

corresponding Tier II assays (Table 4). For example, if a chemical affects only NSI, only those corresponding assays in Tier II would be performed. Since Tier I assays parallel and are predictive of those in Tier II (Luster *et al.*, 1988), this approach provides for efficient cost-effective collection of the most relevant information concerning immunotoxicity.

LOW NATURAL VARIABILITY

Low natural variability is a critical characteristic of a good indicator measurement endpoint. We have discovered that the following crucial coelomocyte parameters and functions are inherently stable in unexposed control earthworms obtained and assayed at various seasons over the past 2 y: (1) Total coelomocyte counts; (2) differential coelomocyte counts; (3) coelomocyte viability; (4) formation of erythrocyte (ER) and secretory (SR) rosettes; and (5) reduction of nitroblue tetrazolium (NBT) dye (e.g. Eyambe *et al.*, 1991; Chen *et al.*, 1991). Additionally, *L. terrestris* has shown consistent response to the 48-h 2-chloroacetamide reference toxicity test (our unpublished data).

Application of earthworm assays to HWS soils

We report here for the first time, results of a preliminary assessment of sublethal toxicity of soils from a Superfund HWS using several of our candidate earthworm assays. This work was in collaboration with Dr. C.A. Callahan, who presents his work on the HWS elsewhere in this publication (also see Callahan *et al.*, 1991). Live *L. terrestris*, exposed in situ to soils having different contamination levels, and soil samples were sent by Dr. Callahan to our laboratories for immunoassays and chemical residue analyses. Immunoassays on in situ-exposed earthworms were conducted blindly (i.e. without knowing the level of in situ exposure toxicity). After assays were completed we categorized earthworms as having been exposed in situ to high, intermediate or low contamination levels according to Dr. Callahan's in situ mortality and morbidity data. These categories corresponded to HWS locations FSS-W1, W2 and W3, respectively (Callahan *et al.*, 1991). We also studied immunologic and pathologic effects in *L. terrestris* exposed in our laboratories to DB-3 soil (Callahan *et al.*, 1991) which was diluted with artificial soil (Greene *et al.*, 1989) to 5%.

GROSS PATHOLOGY

Callahan *et al.* (1991) reported that field exposure produced various gross pathologies. We exposed *L. terrestris* to 5% DB-3 soil for 10, 20 and 30 d, and categorized resultant pathology into five morbidity indicators: Coiling, constriction, edema, evisceration, reddening, ulceration and blistering (Figure 1). Although earthworms showed considerable external pathology, mortality was within expected control limits for the test period (i.e. < 20%). Coiling occurred in over 90% of the worms within the first 10 d. Other indicators appeared at lower frequencies and increased with duration of

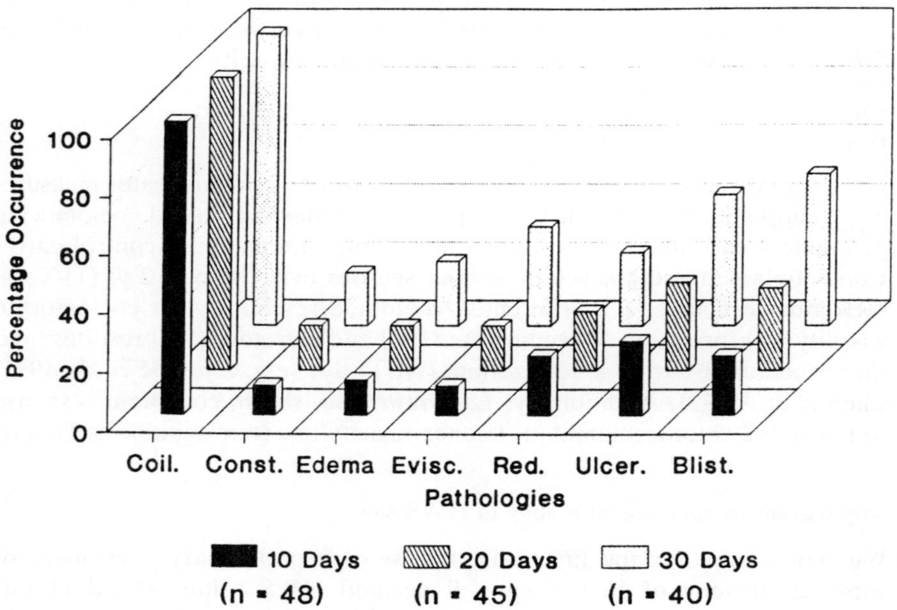

Figure 1. Relation between external morphopathological changes (% occurrence) and exposure time (d) in earthworms, *Lumbricus terrestris*, exposed to hazardous waste site (HWS) soil diluted with artificial soil to 5%. Sample size is given in parenthesis.

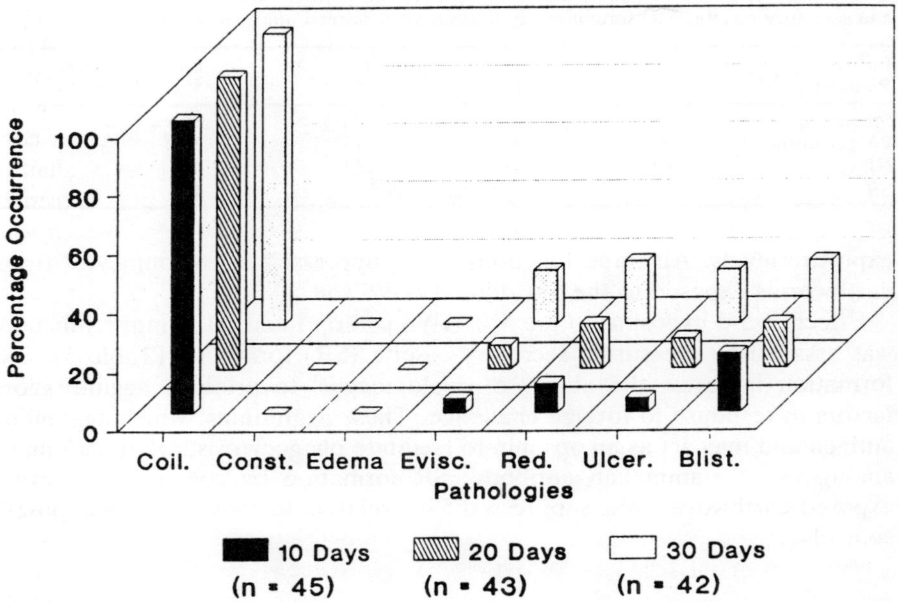

Figure 2. Relation between morphopathological changes (% occurrence) and exposure time (d) in earthworms *Lumbricus terrestris*, exposed to 25·0 mg/kg of chlordane in artificial soil. Sample size is given in parenthesis.

exposure. This pattern was similar to the one we found in *L. terrestris* exposed for 30 d to 25 mg/kg technical chlordane in artificial soil (Figure 2). Chlordane, which was the dominant chlorinated pesticide in the HWS soil (Callahan *et al.*, 1991), was about 25 mg/kg in the 5% DB-3 soil.

IMMUNOTOXICITY

Coelomocyte spreading, phagocytosis of antigenic rabbit red blood cells (RRBCs) and reduction of nitroblue tetrazolium (NBT) dye were used as assays for non-specific immune function (Table 5). Spreading indicates membrane activation, leading to phagocytosis or enhanced ingestion of particulate antigens. NBT reduction indirectly measures the ability of cells to kill ingested microorganisms through production of superoxide and hydrogen peroxide. Reduction in any of these would indicate suppression of non-specific immune function. Assays were performed on coelomocytes from earthworms exposed in situ for 48 h to HWS soils of low, intermediate and high levels of contamination, and from those exposed in the laboratory for 5 d to a 5% dilution of the DB-3 soil. In situ exposure suppressed coelomocyte activation and ability to reduce NBT at all three levels. Phagocytosis was reduced at intermediate and high levels. There was an apparent 'dose response' ranging from 13 to 100% of normal values from high to low

Table 5. Effects of in situ exposure to hazardous waste site (HWS) soil of low (L), intermediate (I) and high (H) levels of contamination, and laboratory exposure to 5% dilution of HWS soil on coelomocyte activation (spreading), phagocytosis, nitroblue tetrazolium (NBT) dye reduction and secretory rosette (SR) formation. Expressed as % normal unexposed controls.

Immune Parameter	In-situ Exposure L	I	H	Laboratory Exposure 5% HWS Soil
Spreading	76	22	45	67
Phagocytosis	100	44	13	28
NBT	67	43	13	45
SR		– Not determined –		47

exposure levels. All three functions were suppressed in coelomocytes from earthworms exposed to the 5% diluted HWS soil.

Effect of 5-d exposure to the 5% HWS soil on humoral immune function was assayed by measuring secretory rosette (SR) formation (Table 5). SR formation indicates the ability of coelomocytes to produce agglutination factors in response to foreign challenge. These agglutinins, which aggregate antigen and may act as an opsonin to facilitate phagocytosis, are functionally analogous to mammalian antibody. SR formation by coelomocytes from exposed earthworms was suppressed 53% relative to those from unexposed controls.

Conclusions

Earthworms have considerably greater potential for use in environmental toxicology than generally realized. They are sufficiently complex with systems and processes that are broadly conserved phylogenetically for use as surrogates for vertebrates in studying environmental toxicants. A number of coelomocyte-based endpoints have been shown to be sensitive indicators of sublethal toxicity of single chemicals and complex mixtures. Preliminary application of indicators of morbidity and immune suppression to HWS soil demonstrates good correspondence between field and laboratory exposures. Development of a comprehensive suite of indicator measurement endpoints, representing various levels of integration within a tiered framework should provide for cost-effective assessment of potential risks to both humans and wildlife from terrestrial pollution.

Acknowledgements

Our research has been supported by grants from the National Institute of Environmental Health Sciences (#ES34811-01) and the Texas Advanced Technology Program. Our special thanks go to Dr. Clarence A. Callahan for his collaboration and support.

References

CALLAHAN, C.A., MENZIE, G.A., WILBORN, D.C., ERNST, T. (1991). On-Site methods for assessing chemical impact on the soil environment using earthworms; A case study at the Baird & McGuire Superfund Site, Holbrook, MA. *Environmental Toxicology and Chemistry*, **10**, 817–826.

CHEN, S.C., FITZPATRICK, L.C., GOVEN, A.J., VENABLES, B.J., AND COOPER, E.L. (1991). Nitroblue tetrazolium dye reduction by earthworm *Lumbricus terrestris* coelomocytes: An enzyme assay for non-specific immunotoxicity of xenobiotics. *Environmental Toxicology and Chemistry*, **10**, 1037–1043.

DEAN, J.H., LUSTER, M.I., BOORMAN, G.A., AND LAUER, L.D. (1982). Procedures available to examine the immunotoxicity of chemicals and drugs. *Pharmacology Review*, **34**, 137–148.

DEAN, J.H., LUSTER, M.I., MUNSON, A.E., AND AMOS, H. (EDS) (1985). *Immunotoxicology and Immunopharmacology*. Raven Press, New York, pp. 1–510.

DEAN, J.H., MURRAY, M.J., AND WARD, E.C. (1986) Toxic responses of the immune system. In: *Casarett and Doull's Toxicology*, C.D. Klaassen, M.D. Andur, and J. Doull, (Eds), Macmillan, New York, pp. 245–285.

EYAMBE, G., GOVEN, A.J., FITZPATRICK, L.C., VENABLES, B.J., AND COOPER, E.L. (1991). Extrusion protocol for use in chronic immunotoxicity studies with earthworm *Lumbricus terrestris* coelomic leukocytes. *Laboratory Animals*, **25**, 70–76.

FITZPATRICK, L.C., GOVEN, A.J., VENABLES, B.J., AND COOPER, E.L. (1990). Earthworm immunoassays for evaluating biological effects of exposure to hazardous materials. In: *In Situ Evaluation of Biological Hazards of Environmental Pollutants*, S.S. Sandhu (Ed) Plenum Press, NY, pp. 119–129.

FITZPATRICK, L.C., SASSANI, R., VENABLES, B.J., AND GOVEN, A.J. (1992). Comparative toxicity of polychlorinated biphenyls to earthworms *Eisenia foetida* and *Lumbricus terrestris* (In Press *Environmental Pollution*).

GOVEN, A.J., VENABLES, B.J., FITZPATRICK, L.C., AND COOPER, E.L. (1988). An invertebrate model for analyzing effects of environmental xenobiotics on immunity. *Clinical Ecology* **4**, 150–154.

GREENE, J.C., BARTELS, C.L., WARREN-HICKS, W.J., PARKHURST, B.R., LINDER, G.L., PETERSON, S.A., AND MILLER, W.E. (1989). *Protocols for short-term toxicity screening of hazardous waste sites*. U.S. Environmental Protection Agency. EPA/600/3–88/029. ERLC. Corvallis, OR.

LUSTER, M.I., MUNSON, A.E., THOMAS, P.T., HOLSAPPLE, M.P., FENTERS, J.D., WHITE, K.L., LAVER, L.D., GEMOLEC, D.R., ROSENTHAL, G.J., AND DEAN, J.H. (1988). Methods evaluation: Development of a testing battery to assess chemical-induced immunotoxicity: National Toxicology Program's guidelines for immunotoxicity evaluation in mice. *Fundamentals of Applied Toxicology*, **10**, 2–19.

RODRIGUEZ, J., VENABLES, B.J., FITZPATRICK, L.C., GOVEN, A.J., AND COOPER, E.L. (1989). Suppression of secretory rosette formation by PCBs in *Lumbricus terrestris*: An earthworm immunoassay for humoral immunotoxicity of xenobiotics. *Environmental Toxicology and Chemistry*, **8**, 1201–1207.

SHARMA, R.P. (ED.). (1981) *Immunologic considerations in toxicology*. Vols. 1 & 2, CRC Press, Inc., Boca Raton, Florida.

SUTER II, G.W. (1989) Ecological endpoints. In *Ecological Assessment of Hazardous Waste Sites: A field and laboratory reference*. W. Warren-Hicks. B.R. Parkhurst and S.S. Baker, Jr. (Eds), EPA/600/3–89/013, pp. 2.1–2.26.

VENABLES, B.J., DAUGHERTY, K.E., GOVEN, A.J., AND OHLSSON, O.O. (1988). Characterization of toxicity of fly ash from the combustion of refuse-derived fuel. *Proceedings of the 81st Annual Meeting of the Air Pollution Control Association, Dallas, TX,* **88–26**, p. 3–16.

VOS, J.G. (1977). Immune suppression as related to toxicology. *CRC Critical Review of Toxicology*, **5**, 67–101.

VOS, J.G. (1980). Immunotoxicity assessment: Screening and function studies. *Archives of Toxicology* Suppl., **4**, 95–108.

WOJDANI, A., STEIN, E.A., AND COOPER, E.L. (1982). Agglutinins and proteins in the earthworm, *Lumbricus terrestris*, before and after injection of erythrocytes, carbohydrates and other materials. *Development Comparative Immunology*, **6**, 612–624.

Poster papers

Experiences and Problems with the OECD-Earthworm Acute Test in Routine Testing

J. RÖMBKE, P. VICKUS AND C. BAUER

Dept. Toxicology & Pharmacology, Battelle Institute e.V., D-6000 Frankfurt 90, Germany

Introduction

Numerous papers concerning laboratory studies to determine the toxicity of chemicals, mainly pesticides, to earthworms have been published during the last 10 to 15 years. Few papers, however, have reported results obtained according to the OECD guideline 207 'Earthworm, Acute Toxicity Test' (e.g. Roberts & Dorough 1985). The number of papers on the toxicity to earthworms decreases even more if only studies without any deviation from this guideline are taken into consideration (e.g. Heimbach 1984). In this paper results are summarized which were gained at Battelle when performing earthworm tests according to the OECD guideline routinely. In Germany, this test is required for the registration of pesticides and, dependent on the amount of chemical produced per year, for the notification of chemicals. Additionally, the test has been used in our laboratory to investigate the influence of abiotic factors on the toxicity of chemicals to earthworms.

Test conditions and parameters analyzed

All data presented here are based on some 40 tests performed during the period of October 1989 to January 1991. The tests were carried out according to the OECD guideline 207 in compliance with GLP regulations (OECD 1984). The evaluation of the test results were extended to calculate the following parameters:

- in addition to the LC_{50} the No-Observed-Effect-Concentration ($NOEC_B$) for effects on the biomass was determined;
- the five concentration levels usually required were graded by a factor of 1·73 (= $\sqrt{3}$); in case this was not sufficient to determine the $NOEC_B$, more concentration levels were tested;
- to calculate the LC_{50} Probit Analysis and Probit Transformation were applied; Analysis of Variance (ANOVA) was used to differentiate between

the $NOEC_B$ and the $LOEC_B$ (Lowest-Observed-Effect-Concentration on biomass) ($P = 0{\cdot}05$).

Experiences

The test substances applied can be characterized as fungicides (33%), herbicides (25%), insecticides (20%) (including microbial pesticides) and other chemicals (22%). The following data were obtained:

- According to Edwards (1984) and others (e.g. Viswanathan & Korte 1984) the LC_{50} value for the reference substance Chloroacetamide should range between 20–40 mg/kg. In our laboratory the LC_{50} value for Chloroacaetamide was on average $19{\cdot}6 \pm 11{\cdot}9$ mg/kg ($n = 5$).
- The tests were performed in compliance with the quality criterion that the mortality of the control organisms should be less than 10% at the end of the test (Tab. 1). Even in field soils (e.g. a sandy loam) *E. fetida* can survive for at least 28 days. More than 50% of all test substances investigated caused no mortality up to 1000 mg/kg.
- Another quality criterion recommended to be added to the guideline was the reduction of biomass for the control organims by less than 10% during the test period of 14 days (Tab. 1). In a sandy loam the weight was 97·3% after 14 days and 87·5% after 28 days.
- If the Bartlett's Test indicated homogeneity of the test results concerning biomass, the ANOVA revealed statistically significant differences between treatment levels and the control when on average the biomass changes were approximately 15–20%.
- Test substances could induce reductions in biomass of up to 30% without causing mortalities (Fig. 1). In some cases worms survived the loss of 60% of their initial weight during the test period. In general, however, a loss of approximately 35% biomass would lead to mortality in an extended test period.
- Surprisingly, a microbial pesticide caused a statistically significant increase of the biomass. At a test substance concentration of 1000 mg/kg the biomass was 112% of the initial weight after 14 days.
- In 40% of all tests both the LC_{50} and the $NOEC_B$ were higher than 1000 mg/kg; in 20% of all tests the LC_{50} was higher than 1000 mg/kg but the $NOEC_B$ lower; in all other cases both values were below 1000 mg/kg. In the latter group the ratio between the two test parameters varied considerably: the $NOEC_B$ ranged between 3 and 97% of the LC_{50}.
- The highest toxicity was determined for insecticides ($LC_{50} = 311$ mg/kg), second became the fungicides ($LC_{50} = 713$ mg/kg), third the herbicides ($LC_{50} = 763$ mg/kg). For fungicides the $NOEC_B$ was considerably lower (396 mg/kg) than the LC_{50} value. These effect values were 100 times or more higher than the recommended field application rates for the pesticides. Other chemicals tested so far showed, in general, a very low toxicity to earthworms.

Table 1. Mean and Standard Deviation (% of the initial number) of the mortality and the biomass of the earthworms in untreated Artificial Soil (n = 40) after a test period of 7 and 14 days

Test	7 Days	14 Days
Mortality		
Range Finding Test	2.30 ± 5.60	3.50 ± 6.10
Definitive Test	2.15 ± 3.10	5.60 ± 3.25
Biomass		
Range Finding Test	98.7 ± 7.2	92.7 ± 7.4
Definitive Test	100.7 ± 6.4	94.6 ± 5.8

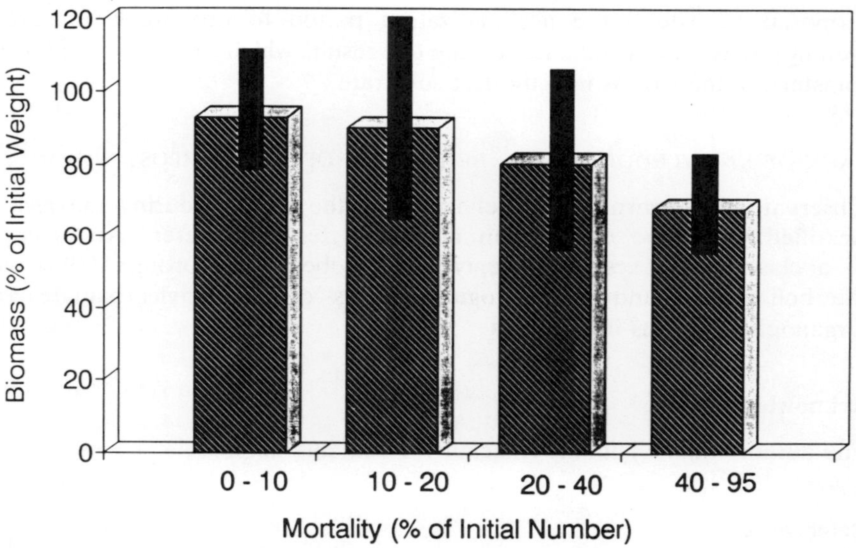

Figure 1. Relationship between mortality and biomass (n = 40) shown as the average weight loss (incl. standard deviation) in different mortality classes

Problems

VARIABILITY OF HOLDING AND TEST CONDITIONS

Under holding conditions the reproduction rate of *E. fetida* occasionally dropped drastically. This is thought to be possibly caused by:

- inadequate food quality (dried cow dung can have harmful ingredients or can be over dried). Buying cow dung from a farm where no pesticides are used might help to avoid this problem.
- inadequate peat treatment (too finely ground peat is not suitable for *E. fetida*).
- variability of sphagnum peat qualities (for example, the maximum water capacity of the Artificial Soil can fluctuates between 60 and 120%).
- breeding of *E. fetida* could certainly be improved if more would be known on the many different parasites which can effect the test organisms (e.g. Meier 1955).

Based on experiences with two chemicals Heimbach & Edwards (1983) stated that the composition and properties of the Artificial Soil had little influence on the test results. Despite the fact that these investigations should be repeated with more compounds it is the sphagnum peat which seems to cause most of the variability occurring in the test.

ACCLIMATIZATION PERIOD

The guideline requires an acclimatization period in Artificial Soil for the test animals of one day. In agreement with van Leemput *et al.* (1989), it is proposed to extend the acclimatization period to one week (including feeding) to avoid any influence on the test results which can be induced when transferring the worms into the test substrate.

LACK OF KNOWLEDGE ON THE BEHAVIOUR OF THE COMPOST WORM

Observations concerning the behaviour of the animals during the test, if classified, could be used as an additional test parameter. An example of a classification system was given by Roberts & Dorough (1984) for the behavioural and morphological effects of N-Methylcarbamate and Organophosphorous insecticides.

Acknowledgements

The authors thank Thomas Knacker for valuable help.

References

EDWARDS, C.A. (1984). Report of the second stage in development of a standardized laboratory method for assessing the toxicity of chemical substances to earthworms. Commission of the European Communities. EUR 9360 EN, 99 pp.

HEIMBACH, F., EDWARDS, P.J. (1983). The toxicity of 2–Chloroacetamide and Benomyl to earthworms under various test conditions in an Artificial Soil Test. *Pesticide Science* **14**, 635–636.

HEIMBACH, F. (1984). Correlations betwen three methods for determining the toxicity of chemicals to earthworms. *Pesticide Science* **15**, 605–611.

MEIER, M. (1955). Parasitische Ciliaten bei Oligochäten. *Archiv für Protistenkunde* **100**, 212–245.

ROBERTS, B.L., DOROUGH, H.W. (1984). Relative toxicities of chemicals to the earthworm *Eisenia fetida. Environmental Toxicology and Chemistry* **3**, 67–78.

VAN LEEMPUT, L., SWYSEN, E., WOESTENBORGHS, R., MICHIELSEN, L., MEULDERMANS, W., HEYKANTS, J. (1989). On the terrestrial toxicity of the fungicide Imazalil (Enilconazole) to the earthworm species *Eisenia fetida. Ecotoxicology and Environmental Safety* **18**, 313–320.

VISWANATHAN, R., KORTE, F. (1984). A laboratory study to determine the toxicity of 15 organic chemicals to Eisenia fetida using an artificial medium. In: Actes Symposium Internationale 'Ecotoxicologie terrestre', Les Arcs, 1984. pp. 629–636.

A Laboratory Test on Sublethal Effects of Pesticides on *Eisenia fetida*

C. KOKTA

Federal Biological Research Centre for Agriculture and Forestry, Department for Plant Protection Products and Application Techniques, Zoology Division, Messeweg 11/12, D-3300 Braunschweig, Germany

Abstract

This paper describes a method for testing side effects of pesticides on reproduction and body weight of *Eisenia fetida*. The results of a ringtest with two pesticides show that this test design might be useful within a stepwise testing procedure on effects for pesticides on earthworms.

Introduction

For the investigation of side effects of chemicals on earthworms in the laboratory, tests on acute toxicity are well established (e.g. OECD-guideline No. 207). However, a laboratory test with special regard to sublethal effects that comprises an exposure scheme according to the ordinary field use of the pesticide is lacking. This paper presents a new method and results of a ringtest with two pesticides.

Method

The test design is in some points closely related to OECD-guideline No. 207 (OECD, 1984). The same artificial soil substrate is used. Adult *Eisenia fetida fetida* or *Eisenia fetida andrei* from a synchronized laboratory culture are introduced into the test boxes.

Plastic test boxes with a surface area of 200 cm^2 and a transparent cover lid were used by all participants. About 600 g dry weight of artificial soil were filled into each test box. This amount corresponds to a substrate height of 5–6 cm. Ten animals per test box with 3–4 replicates per concentration were used.

The pesticide was applied as close as possible to field conditions. The amount of water used for application corresponded to 600–800 l/ha. Two pesticides (Du Pont Benomyl, a.i. benomyl, and E 605 forte, a.i. parathion) were sprayed with the recommended dose and a 5-fold dose. With benomyl the normal dose was 250 g formulated product/ha (= 125 g a.i./ha) and with parathion it was 210 ml formulated product/ha (= 105 1 ml a.i./ha).

Food was added to ensure sufficient reproduction. Finely ground cow manure from the same charge was used by all laboratories. Before the beginning of the test an amount of 0·5 g per animal was mixed into the soil homogenously. After the application of the pesticide 0·5 g per animal was spread on the soil surface once a week and moistened. To avoid food going mouldy, it was only given when feeding activity was observed. During the test, the test boxes were stored at 20 ± 1°C at a light to dark cycle of 16:8.

The adult earthworms were exposed for 6 weeks. After this time the alive adults were removed from the substrate and counted. Additionally, the body weight was determined. The offspring stayed in the test boxes for another 4 weeks to ensure hatching of cocoons.

Food was given when required. After 10 weeks the juveniles were removed from the substrate by handsorting or heat extraction and counted.

Table 1. Effects of benomyl on the number of juveniles and on the body weight of *Eisenia fetida* (1-fold application rate: 250 g/ha; mean out of 3–4 replicates and standard deviation; * = significant different from control, p › 0·05, Tukey-test)

Laboratory	Control	1-fold Appl.	5-fold Appl.
% Mortality			
1	0	0	3 ± 6
2	15 ± 19	5 ± 10	0
3	0	3 ± 6	0
4	0	0	0
5	0	0	0
6 (after 4 weeks)	7 ± 6	13 ± 15	10 ± 10
7	0	3 ± 6	27 ± 38
8 #	0	0	0
Number of juveniles/test box after 10 weeks			
1	38 ± 16	17 ± 19	0·3 ± 0·5*
2	88 ± 17	95 ± 14	50 ± 8*
3	36 ± 4	25 ± 12	17 ± 14
4	53 ± 10	57 ± 14	3 ± 3*
5	85 ± 61	121 ± 11	62 ± 25
6 (after 8 weeks)	29 ± 16	19 ± 11	9 ± 2
7	31 ± 4	14 ± 2*	1 ± 1*
8 #	120 ± 41	69 ± 25*	17 ± 3*
Weight of adults after 6 weeks, % of initial weight			
1	156 ± 9	155 ± 19	66 ± 5*
2	125 ± 7	127 ± 5	130 ± 7
3	90 ± 6	97 ± 8	98 ± 3
4	143 ± 12	140 ± 7	147 ± 14
5	261 ± 15	245 ± 29	213 ± 32
6 (after 4 weeks)	116 ± 4	94 ± 9*	93 ± 5*
7	121 ± 15	101 ± 12	53 ± 33*
8 #	154 ± 2	115 ± 7*	76 ± 2*

In this laboratory the 1·5-fold resp. 7·5-fold dose was tested

Table 2. Effects of parathion on the number of juveniles and on the body weight of *Eisenia fetida* (1-fold application rate: 210 ml/ha; mean out of 3–4 replicates and standard deviation; * = significant different from control, p ‹ 0·05, Tukey-test)

Laboratory	Control	1-fold Appl.	5-fold Appl.
% Mortality	1	0	0
2	0	0	0
3	0	0	0
4	10 ± 14	5 ± 10	5 ± 10
5	0	0	0
6 (after 8 weeks)	7 ± 6	17 ± 12	7 ± 12
7	0	0	0
8 #	0	0	0
Number of juveniles/test box after 10 weeks			
1	38 ± 16	18 ± 17	11 ± 8
2	36 ± 4	23 ± 7*	22 ± 1*
3	53 ± 10	76 ± 27	45 ± 23
4	148 ± 31	131 ± 15	89 ± 22*
5	94 ± 10	73 ± 15	100 ± 60
6 (after 8 weeks)	29 ± 16	39 ± 5	22 ± 3
7	31 ± 4	20 ± 3*	12 ± 2*
8 #	149 ± 28	140 ± 15	96 ± 31*
Weight of adults after 6 weeks; % of initial weight			
1	156 ± 9	175 ± 19	168 ± 15
2	90 ± 6	96 ± 9	82 ± 4
3	143 ± 12	135 ± 6	73 ± 1*
4	86 ± 12	92 ± 11	93 ± 9
5	234 ± 4	270 ± 16	254 ± 50
6 (after 4 weeks)	116 ± 4	98 ± 6*	97 ± 4*
7	121 ± 2	115 ± 6	93 ± 10*
8 #	146 ± 19	143 ± 7	138 ± 18

In this laboratory the 1·5-fold resp. 7·5-fold dose was tested

Participants of the ringtest 1990/91

P. Dohmen; BASF AG (FRG)
R. Fischer; Hoechst AG (FRG)
F. Heimbach; Bayer AG (FRG)
C. Kokta; Federal Research Centre for Agriculture and Forestry (FRG)
H. Kula; TU Braunschweig (FRG)
F. Riepert; Federal Resarch Centre for Agriculture and Forestry (FRG)
J. Römbke; Battelle Institute, Frankfurt (FRG)
U.Thielemann; GEFAÖ, Nußloch/Heidelberg (FRG)
C.A.M. Van Gestel; National Institute of Public Health and Environmental Protection (NL)

Results

In most of the tests low mortality of adults occurred in the control as well as in the different treatments. The number of juveniles per adult in the control boxes ranged between 3 and 15 after 10 weeks of exposure.

Du Pont Benomyl had significant effects on juvenile numbers and/or weight in 3 out of 8 tests in the 1-fold rate (table 1). In the 5-fold rate there were significant effects on juvenile numbers in 5 out of 8 tests. Weight was affected in 4 out of 8 tests.

With parathion in the 1-fold rate 2 tests showed significant effects on juvenile numbers. In the 5-fold rate 4 tests resulted in significant effects on juvenile numbers. In the 1-fold rate in 1 test significant effects on weight occurred and in the 5-fold rate 3 tests showed significant effects on weight.

Discussion and outlook

This laboratory test was aimed to get information on sublethal effects on earthworms. The results indicate that the principle of the test is suitable. Nevertheless there are a lot of questions to discuss and further standardization might help to improve the reproducibilty.

The application technique was not the same in the different laboratories and therefore variability may result from this. Because of the different feeding activity the amount of food given was not the same in the tests. The different feeding activity possibly resulted in different fertility. Food was distributed from one charge.

Until now there is no idea what variability in fertility will occur when food from different sources is used. Nevertheless in tests with low numbers of juveniles the same results were found.

Further research and another ringtest is planned to improve standardization and reproducibility.

References

OECD (1984). Guideline for testing of chemicals, No. 207. Earthworm acute toxicity tests. Adopted 4 April 1984.

Study of Pesticides Impact on Earthworms using a Closed Laboratory Model Ecosystem

R. VISWANATHAN*

GSF-Institute of Soil Ecology, 8042 Neuherberg, Germany

Abstract

A closed model ecosystem consisting of a sandy loam soil substrate, maize seedlings and earthworms (*Eisenia fetida* Sav.) in a 30 cm desiccator was used to investigate the impact of soil treated (benzyl U-^{14}C) cypermethrin on earthworms. Exposure to 5·5 mg/kg dry weight of the pesticide for 2 weeks had apparently no adverse effect on earthworms. The worms' tissues had contained 2·6 times ambient concentration of residues and approximately 92% of them were found to be biotransformation products.

Introduction

Different laboratories have employed different methods to elucidate the exposure effects of pesticides and other chemicals on the earthworms. In our laboratory, a closed model ecosystem of R.E. Schroll (1989) was suitably modified and used to investigate many environmental aspects including the impact of soil applied pesticide on the earthworms, *Eisenia fetida* Sav. (Benzyl U-^{14}C) Cypermethrin was used in this model system.

Experimental

The model ecosystem (Set up shown in Fig. 1) was maintained in a phytotron where the average temperature was 20°C, light intensity, approximately 7000 Lux, with a 12 hour photoperiod. Fifteen adult *E. fetida* (average wt. 460 mg) were exposed for 2 weeks in approx. 1200 g sandy loam soil (pH 6·9; moisture content, 18%) which contained 5% cattle manure and which was mixed with 5·5 mg/kg dry weight repurified (benzyl U-^{14}C) cypermethrin

* Part of a work (unpubl.) carried out in collaboration with Mr. Zhen-Hui Zhou (Academia Sinica, Institute of Entomology, Shanghai, Peoples Republic of China).

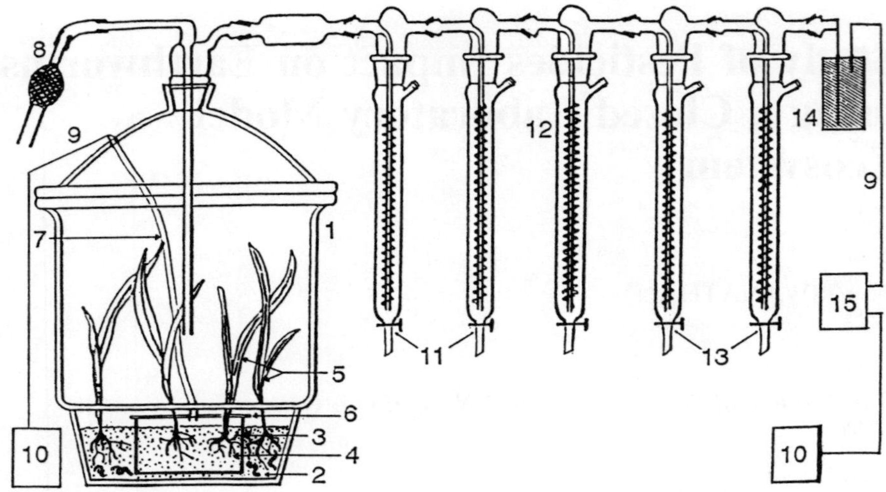

Figure 1. Scheme of closed model ecosystem set up: 1. 30 cm-Desiccator, 2. ^{14}C-Cypermethrin treated soil with *Eisenia fetida* worms, 3. Glass dish, 4. untreated soil, 5. Maize seedlings 6. Teflon disc with holes just big enough to accommodate growing seedlings, 7. Teflon tubing, 8. Adsorption tube with dry silica gel and activated charcoal, 9. Hard PVC tubing, 10. Membrane pump, 11. Traps filled with ethyleneglycol monomethylether plus Permablend (1:1), 12. Empty trap, 13. Traps filled with Carbosorb II plus Permafluor V (2:3), 14. Adsorption tube with activated charcoal, 15. Gas flow rate gauge fitted with fine regulating valve

[0·38 mg (benzyl U-^{14}C) cypermethrin (Shell Research Ltd., U.K.; specific activity, 9·95 mCi/mMol; radiochemical purity, › 99%) + 2·59 mg and 2·53 mg, respectively, unlabelled cis and trans cypermethrin (RIPCORD fractions; purity (thin layer chromatography; GLC – electron capture detector), 94·5% and 97·4%, respectively)] and placed in the desiccator. Eleven one week-old maize seedlings were planted in the treated soil just before start of the experiment. A glass dish containing 500 g untreated sandy loam soil with 4 one week-old maize seedlings was maintained simultaneously in the desiccator for purposes of control. A teflon disc with perforations only to accommodate growing maize seedlings placed on the glass dish additionally helped to protect the untreated soil from direct contamination of the radioactive pesticide.

Fresh and filtered air (approx. 16·5 ml/min) was passed through the whole system continuously. The untreated soil in the control was aerated additionally.

Traps filled, respectively, with ethyleneglycol monomethylether plus Permablend and Carbosorb II plus Permafluor V attached to the desiccator facilitated determination of the volatilised ^{14}C – substance and mineralised $^{14}CO_2$.

Results and discussion

The distribution of the soil applied (benzyl-^{14}C) cypermethrin is shown in Table 1. In 2 weeks the worms accumulated about 2·2% of total applied

Table 1. Distribution of radioactivity in percent after soil application of 5·5 mg/kg dry weight (benzyl-^{14}C) cypermethrin in a closed model ecosystem.

Samples	Treatment			Control		
	Extractable		Unextractable	Extractable		Unextractable
	Organic*	Water		Organic*	Water	
Soil	75·4	0·70	13·5	ND	ND	ND
Earthworm	1·67	0·10	0·38			
Maize						
Roots	0·16	0·22	0·43	0·02	ND	0·05
Stems	0·07	0·06	0·12	ND	ND	ND
Leaves	0·02	0·08	0·25	0·03	0·05	0·09
Air ($^{14}CO_2$)			1·7			

* Successive extraction in benzene, chloroform and methanol
ND Not detectable

Table 2. Degradation/conversion in per cent of (benzyl-^{14}C) cypermethrin in a closed ecosystem in relation to total application

Samples		TLC Fractions						Bound/Conjugated
	Unconverted ^{14}C-cypermethrin	Degradation/conversion products Rf-Values*						residues
		0·00–0·10	0·17	0·30	0·55	0·65	0·83	
Soil	74·1	0·63	0·06	0·23	3·8	0·33		15·5
Earthworm	0·18	0·18	0·25	0·27	0·18	0·12	0·12	0·55
Maize	0·16	0·16			0·05			1·20
Air^{14}(CO_2)					1·7			

* After successive runs in benzene + n-hexane (3+2) 3x and benzene + ethylacetate + methanol (15+15+1) 1x

radioactivity. This meant, considered on the basis of biomass, the worms had in their tissues approx. 2·6 times ambient concentration of the substance. However, the worms were not apparently affected. On the other hand, they had gained 11·2% in weight.

The residues accumulated in the tissues were composed of nearly 92% biotransformation products. By thin layer chromatography 6–7 metabolic components were separated from the organic soluble fraction (Table 2). Two of the components with Rf-values 0·55 and 0·65, respectively, corresponded to 3-phenoxybenzaldehyde and 4′-methoxy-3-phenoxybenzaldehyde.

References

SCHROLL, R.E. (1989). Aufnahme ^{14}C-markierter Chemikalien und Radionuklide aus Boden in Pflanzen in Abhaengigkeit von Pflanzeneigenschaften. *Doctoral thesis*, Technical University Munich, Freising-Weihenstephan, Germany.

The Daniel Funnel Test in a Sequence of Earthworm Tests of Agricultural Pesticides

C.H. HÖGGER[1], H.-U. AMMON[1] AND M. BIERI[2]

[1] *Swiss Federal Research Station for Agronomy, Reckenholz, 8046 Zurich, Switzerland*
[2] *Soil Zoologist and Consultant, Bahnhofstr. 30, 8803 Rüschlikon, Switzerland*

Introduction

The Daniel Funnel Test according to Bieri *et al.* (1989) measures the effect of pesticides on the activity of agriculturally important, vertically tunnelling (anécique) species such as *Lumbricus terrestris* L. This paper presents results obtained with molluscicide baits using the described method.

In addition, we propose a three-step testing sequence, which gradually eliminates 'non-effective' formulations and application methods of pesticides in two laboratory tests, ie. a general toxicity test (Ammon, 1985), a behavioral test (Bieri *et al.*, 1989), and a field test (Edwards & Brown, 1982).

Materials and methods

L. terrestris was collected from permanent grassland after extraction with 0·005% chloroacetophenone in water (Daniel, 1986). Polyethylene funnels of 12 cm diameter were fitted with a silicon rubber tube of 1·2 cm diameter and 30 cm length. The end of the tube was closed with a plug. The tube and the stem of the funnel were darkened with a removable black polyvinyl-chloride pipe. The funnel was filled with non-sterile moist field soil. With a screwdriver the soil was pierced vertically to make an artificial earthworm tunnel. When a juvenile *L. terrestris* of about 2 grams weight was added it retreated into the hole. Five bait granules were put on the surface of the soil in each funnel corresponding to an application rate in a range of 50 to 100 kg bait per hectare. Each treatment was replicated 20 times. The funnel was covered with a mesh screen to prevent the escape of the worms. The funnels were kept in a growth chamber at 12°C for 21 days. After 4, 6, 8, 14 and 21 days the numbers of moved and withdrawn bait kernels were recorded. Then the location of the worm was noted as either in the soil, about in the middle of the tube or at the bottom of the tube. After 21 days the worms were weighed again and watered in a shallow container for 24 h. Then they were

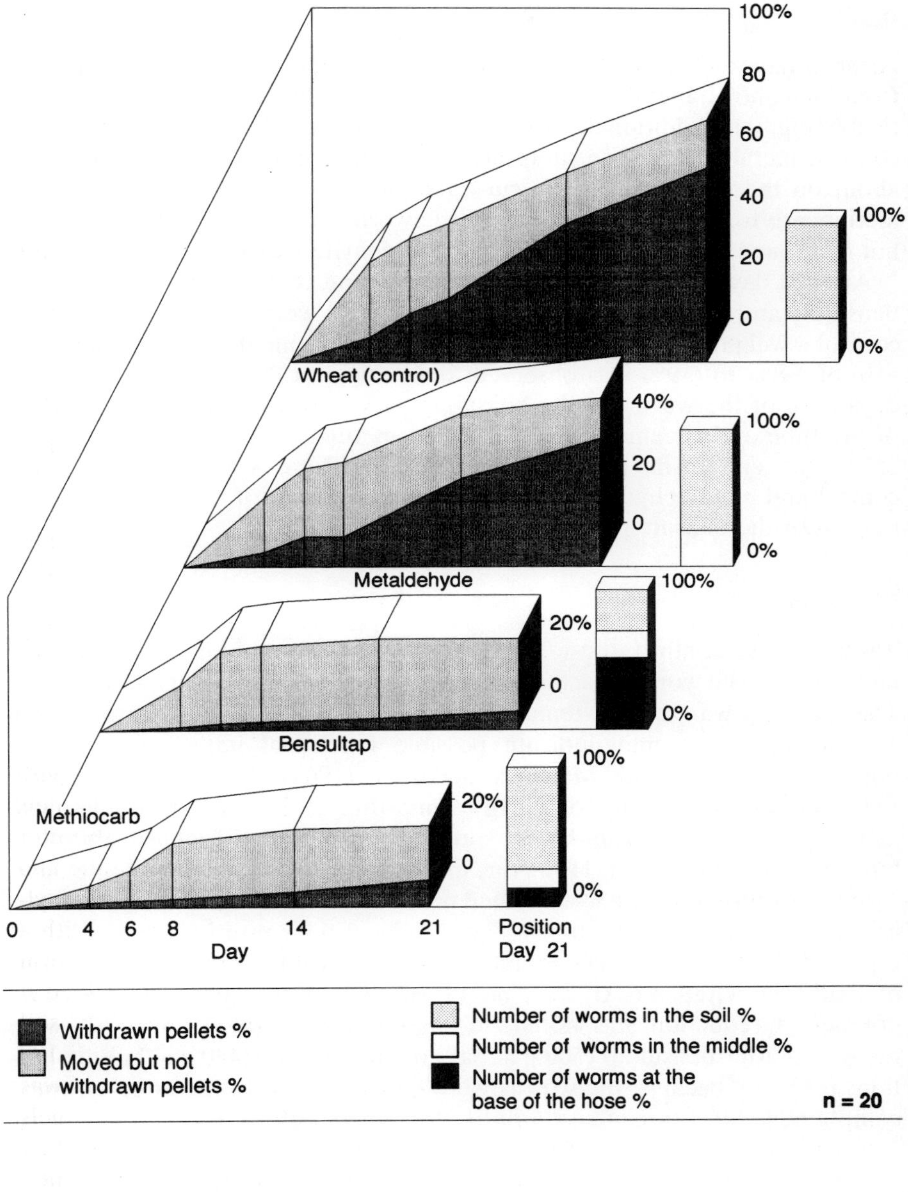
100%
80
60
40
20
0
100%
0%
Wheat (control)
40%
20
0
100%
0%
Metaldehyde
20%
0
100%
0%
Bensultap
Methiocarb
20%
0
100%
0%
0
4
6
8
14
21
Day
Position
Day 21
Withdrawn pellets %
Moved but not
withdrawn pellets %
Number of worms in the soil %
Number of worms in the middle %
Number of worms at the
base of the hose %
n = 20

Figure 1.

weighed again. The test formulations in the Funnel Test were: Limax G: Maag (ai = 3·5% metaldehyde), Malice G: Stauffer, ICI (ai = 5% bensultap) and Mesurol G: Bayer (ai = 1% methiocarb); and as control untreated wheat seed was used.

Results

After 21 days 63% of wheat seed, and 42%, 6% and 9% of baits disappeared from the control, metaldehyde, bensultap and methiocarb treatments respectively (Fig. 1). Additionally 15%, 14%, 23% and 17% respectively of the control, metaldehyde, bensultap and methiocarb granules had been moved about on the soil surface. The curves show also that in the bensultap and methiocarb treatments the activity of the worms ceased after the eighth day, but continued in the control and the metaldehyde treatment.

After 21 days weight losses of 1·1%, 19·1% and 18·3% in the metaldehyde, bensultap and methiocarb treatments respectively were noted, whereas in the control a weight gain of 4·9% was observed. After one day in water a weight gain of 4·8% to 6·9% was observed, depending on the treatment. After 21 days 50% of the worms in the bensultap treatment retreated to the bottom of the tube and became quiescent; in the methiocarb treatment there were 25% quiescent worms at the bottom (Fig. 1). There were no deaths in the control and metaldehyde treatments and only one out of twenty worms died in each of the bensultap and methiocarb treatments.

Discussion

These results confirm those of Bieri *et al.* (1989) on metaldehyde and methiocarb. The worms reacted to bensultap and methiocarb in a similar way. Their activity was at least temporarily reduced and a weight loss occurred. The weight losses may indicate possible functional impairments of the populations in the field. Martin and Forrest (1969) observed a temporary reduction in the activity of earthworms after a 30 second contact with methiocarb granules. *Lumbricus* spp. especially had a reduced ability to burrow into compost soil. However, most worms recovered after some days.

In our experiments the worms lined the hose with castings in all treatments, as they do in natural situations. They changed their position several times, especially in the control. The Funnel Test explains at least partially the results of field tests cited by Atger *et al.* (1990), where an apparent temporary population reduction was noted, ie. significantly fewer *L. terrestris* were recovered after bensultap (500 g ai/ha) and methiocarb (440 g ai/ha) application, probably because the worms were quiescent due to the effect of these compounds. After three to six months the same numbers of *L. terrestris* were extracted from treated and untreated control plots.

The dosage of 5 granules per 113 cm^2 soil surface in the funnel corresponds to about a five to ten times overdose in the field. This dosage was chosen first according to the original method of Brieri *et al.* (1989) and to ascertain the observation of effects, if any. Second, to make this laboratory test more severe than a field test with normal application rates, to allow the exclusion of those non-effective compounds, for which the test sequence need not be continued in the field.

The Daniel Funnel Test is suitable for short-term laboratory studies (3–5 weeks) of the behaviour of anécique earthworms under simulated field conditions, including the influence of pesticides, eg. DNOC (Bieri and Ammon, 1992). Exposition modes can be varied for different forms of pesticide application. Generally, earthworms are fed with wilted clover leaves and food and soil are sprayed and the removal of leaves is the measuring criterion. Variations are: 1) food leaves only may be sprayed; 2) bare soil may be sprayed and no food offered and survival of the worms is the measuring criterion; 3) formulations may be pelleted as in the present example; 4) formulations may be mixed into the soil; and 5) the surface may be flooded after application to wash the formulation into the tunnel to simulate rain storms (Bieri and Ammon, 1992).

At present natural populations of *L. terrestris* are used. Premature juveniles of the same size are taken from old meadows or forests where no pesticides have been applied previously. Fresh catches of worms are aerated in water before use in a test. If more than 5% mortality occurs within three days the batch is rejected. Calibration of the sensitivities of the worm batches may be done by running controls with standard compounds such as chloracetamide.

L. terrestris can be cultured in the laboratory, although reproduction is slow. The generation time is about one year and fertility is low (Lofs-Holmin, 1980). Therefore, it is easier to use juvenile animals from defined wild populations from a non-contaminated forest or grassland. Many problems of standardisation and inbreeding in laboratory cultures are also avoided.

A more standardised version of the funnel test using kaolin china clay as substrate has been accepted by the International Organisation of Biological Control (IOBC) (Bieri, 1991).

References

AMMON, H.U. (1985). Worm toxicity-test using *Tubifex tubifex*. FAO Workshop, Versailles. *Les Colloques de l'INRA* **31**, 303–317.

ATGER, J.C., DELPUECH, I. AND MAURIN, G. (1990). Un nouveau produit antilimaces a base de Bensultap. *ANPP 2ième Conf. Int. Ravageurs Agric., Versailles*. pp. 305–307.

BIERI, M. (1991). Guidelines for the Daniel Funnel Test – a laboratory test to measure side effects of pesticides on the earthworm *Lumbricus terrestris* L. *IOBC Bulletin* (in press).

BIERI, M. AND AMMON, H.U. (1992). Variations in the Daniel Funnel Test for measuring the toxicity of pesticides to earthworms under simulated field conditions. (In preparation).

BIERI, M. AND AMMON, H.U. (1990). Labor- und Freilandmethoden zur Prüfung der Regenwurmgefährdung durch Pestizide. *Landw. Schweiz* **3** (2), 11–15.

BIERI, M., SCHWEIZER, H., CHRISTENSEN, K. AND DANIEL, O. (1989). The effect of metaldehyde and methiocarb slug pellets on *Lumbricus terrestris*. *Slugs and Snails in World Agriculture*. Henderson, I. (Ed.) *BCPC Monograph No. 41*. Thornton Heath: BCPC Publications. pp. 237–244.

DANIEL, O. (1986). Diurnale Effizienzschwankungen beim Sammeln der Regenwurmart *Lumbricus terrestris* mit der Formaldehydmethode. *Bull. Bodenkundl. Ges. Schweiz* **10**, April 1986, 39–44.

EDWARDS, P.J. AND BROWN, S.M. (1982). Use of grassland plots to study the effect of pesticides on earthworms. *Pediobiologia* **24**, 145–150.

LOFS-HOLMIN, A. (1980). Measuring growth of earthworms as a method of testing sublethal toxicity of pesticides. *Swedish J. Agric. Res.* **10**, 25–33.

MARTIN, T.J. AND FORREST, J.D. (1969). Zur Entwicklung von Schneckenkorn Mesurol® in Grossbritannien. *Pflanzenschutznachrichten-Bayer* **22** (2), 212–253.

Relevance of Life-history Parameters in Earthworm Ecotoxicology

D. CLUZEAU[1], R. LAGARDE[2], C. TEXIER[1] AND L. FAYOLLE[3]

[1] *Université de Rennes I, UA CNRS 696 – Lab. Ecol. du Sol & Biol. des Popul., Station Biologique de Paimpont, 35380 Plélan le Grand, France*
[2] *Service de Protection des Végétaux Bretagne, 208 rue de Fougères, 35700 Rennes, France*
[3] *INRA, Station de recherche sur la Flore pathogène et la Faune du Sol, 17 rue Sully, 21034 Dijon Cédex, France*

Abstract

The rationale of the present study, focused on sublethal effects influencing earthworm population dynamics, was to provide an improved basis for analyzing the results of subsequent field studies.

Introduction – study design

In recent years, agronomists have been concerned with the indirect effects of active products on non-target species.

In order to investigate pesticides effects, appropriate methods for classifying active products as a function of toxicity are required (Fig. 1). Determining such methods is difficult, given the diverse criteria used by different authors to evaluate the effects of pesticide treatments on soil fauna. Under experimental and controlled conditions, it is possible to monitor the effects of chemical disturbances on demographic parameters directly influencing the medium-term rate of population growth.

The laboratory study was implemented in accordance with the following assessment levels: (1) *Lethal (direct) effects*, evaluated by immediate changes in survival rate; (2) *Sublethal (indirect) effects*, including the effects of chronic exposure, which may affect the homeostasis of individual organisms, hence population dynamics.

Laboratory studies utilizing multiple assessment criteria at several levels of population dynamics serve to enhance the understanding of subsequent field study results.

The objectives of an overail systems approach are only attainable after

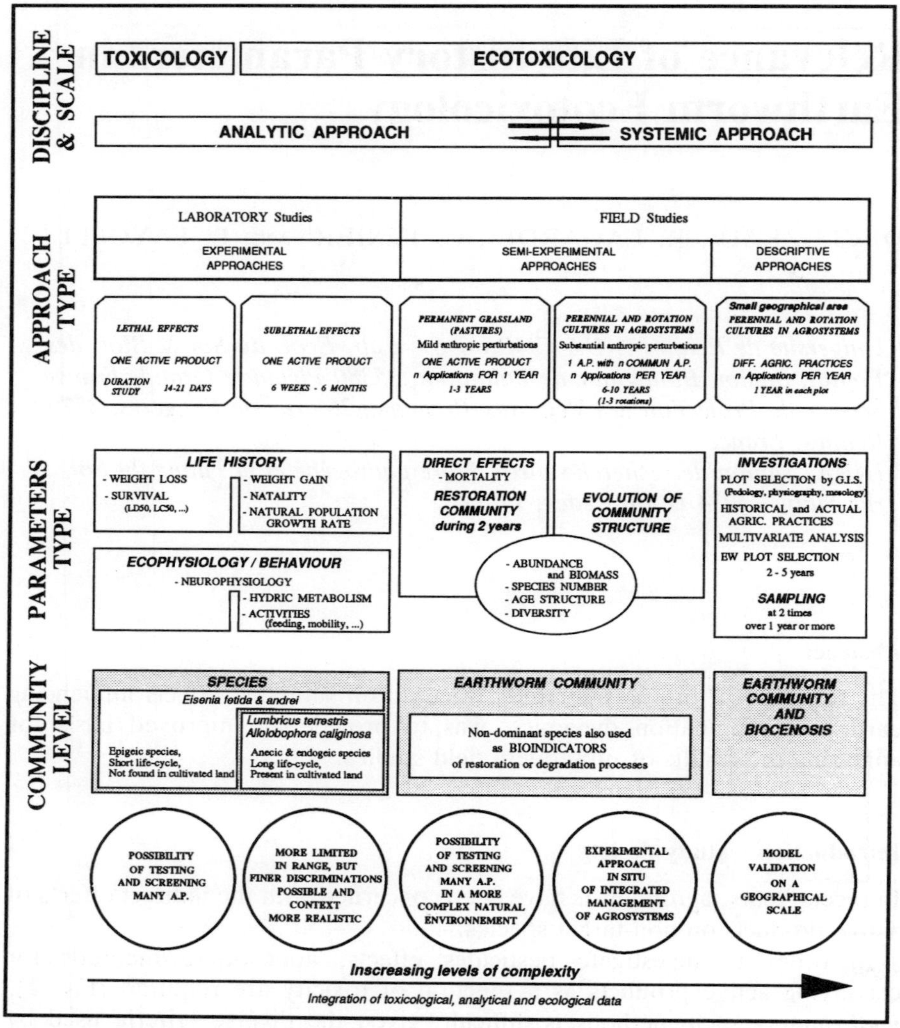

Figure 1. Levels of assessment in earthworm toxicology
(A.P. = Active Product; G.I.S. = Geographical Information System)

laboratory and field data from more circumscribed approaches have been integrated. Although difficult to achieve, synthesis of findings from different levels of research is essential for an adequate understanding of the actual functioning of agrosystem constantly subjected to chemical treatments. The resulting agrosystem models must then be applied to small agricultural areas, in order to validate them for use in the integrated management of real agrosystems.

It is essential to pursue research on both the individual components and mechanisms of an agroecosystem and on the agroecosystem as a whole, as the two types of research serve to validate and expand each other's findings.

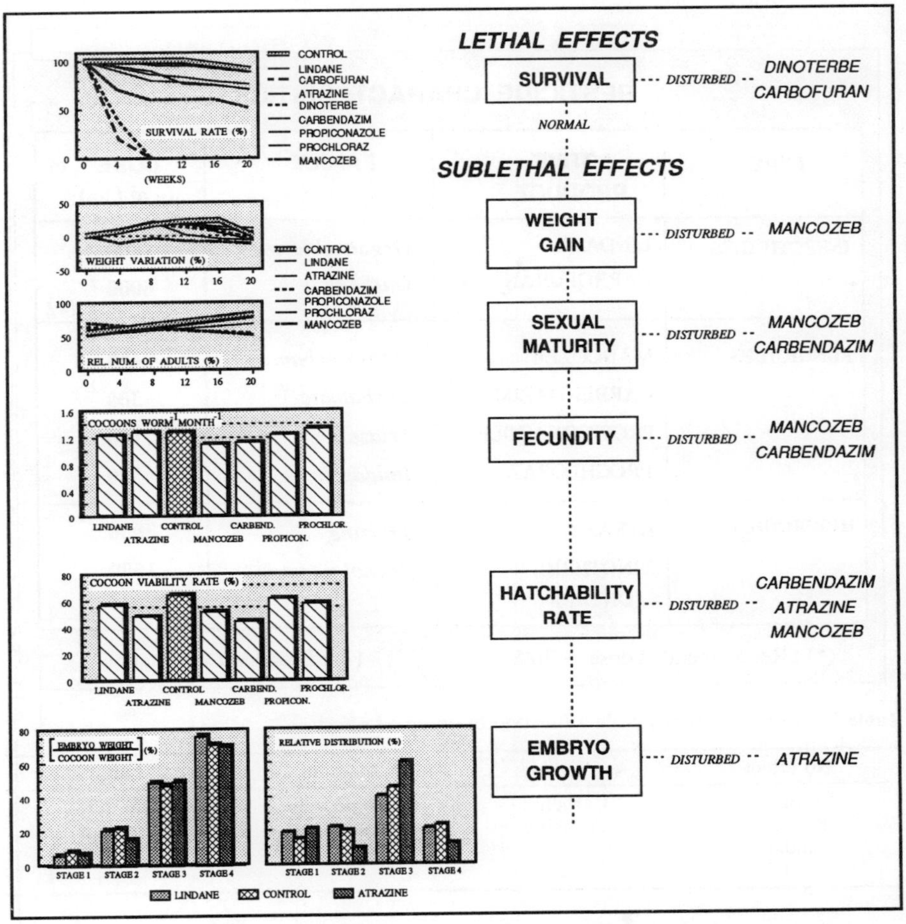

Figure 2. Levels of assessment of life-history parameters in laboratory studies

This dual approach, involving research at increasing levels of complexity, is currently being developed with regard to vineyards of Champagne (Cluzeau *et al.*, 1990).

At present, our subject is limited to evaluating the relevance of biodemographic parameters for assessing the earthworm ecotoxicity of pesticides.

Results and discussion

Results have synthetic presentation to determine the possibility of testing and screening by using life-history parameters (Fig. 2) (for detailed results, cf. Cluzeau et Fayolle, 1990)

The toxicity of the main pesticides utilized in Brittany, France – dinoterbe, atrazine, lindane, carbofuran, mancozeb, carbendazim, prochloraz and propiconazole – were investigated with respect to population parameters of

Table 1. Product characteristics and doses utilized in the study

PESTICIDE CHARACTERISTICS

TYPE	ACTIVE PRODUCT	FAMILY	DOSE (*) (g ai / ha)
INSECTICIDES	LINDANE	*Organochlorine*	1800
	CARBOFURAN	*Carbamate*	6000 (1)
FUNGICIDES	MANCOZEB	*Dithiocarbam.*	2400
	CARBENDAZIM	*Carbamate*	300
	PROPICONAZOLE	*Triazole*	150
	PROCHLORAZ	*Imidazole*	540
HERBICIDES	ATRAZINE	*Triazine*	3000
	DINOTERBE (DNTBP)	*Phenol nitrate*	1500

(*) : Recommended dose + 20% (1) : Broadcast application

Table 2. Relative toxicity scale a function of observed effects

No effect	Weak	Medium	Large
Propicon. Prochlo. Lindane	Carbend. Atrazine	Mancozeb	Dinoterb Carbofur.
0	+	++	+++

Lumbricus terrestris and offspring under controlled laboratory conditions. The rates of application used in this study are presented in table 1.

Products with lethal effects may be initially screened on the basis of survival rates, then further graded according to persistence of biotoxicity. Chronic toxicity (bioaccumulation phenomena and the effects of degradation products), which may effect populations in the field, were evaluated from sublethal effects. Four groups of active products were distinguished, corresponding to a four-point scale of toxicity (Table 2). Except for Group I (maximum toxicity), the toxicity scale values are only relative.

Ecotoxicological studies including detailed parameters on population dynamics provide relevant criteria for evaluating lethal and especially sublethal effects. The results of the present study confirm previous findings that disturbances in non-target population dynamics induced by the repeated application of pesticides classed as only moderately toxic (groups II and III) must be taken into account to prevent interference with the quality and rate of renewal of cultivated soils. Furthermore, products with low toxicity (group

IV) but showing long-term persistence in soils (more than 3 years, in the case of lindane – PETZ and FOISSNER, 1989) can constitute a permanent stress.

Integrated pest control requires taking a broad range of factors into account. Thus, products considered to be effective vis-à-vis target populations but fairly innocuous vis-à-vis non-target ones may subsequently reveal substantial toxicity if their persistence in the environment is high. Bioaccumulation and contaminant transfer in food chains are additional factors which cannot be ignored.

For the above reasons, there is a pressing need for refined, interdisciplinary research on the medium- and long-term toxicity of the main new agropharmaceutical compounds to determine their role in rational management of the rural environment.

Acknowledgements

The authors thank Ms. G. Chaminade and Mrs. Y. Picard for their technical assistance.

References

CLUZEAU, D., AND FAYOLLE, L. (1990). Approche démographique d'une population de *Lumbricus terrestris*, en liaison avec des produits agropharmaceutiques utilisés en polyculture-élevage. Coll. Int. ANPP *'Relations entre les traitements phytosanitaires et la reproduction des animaux'*, Paris, France, 25–26/04/90. pp. 245–256.

CLUZEAU, D., GUO Z.T., CHAUSSOD R., FEDOROF N., ESMENJAUD D., NORMAND M. AND VALENTIN, G. (1990). Etats structuraux et biologiques des sols du vignoble de Champagne. Rapport scientifique, Université de Rennes I, 2 vol., 124 p. & 110 p.

PETZ, W. AND FOISSNER, W. (1989). The effects of mancozeb and lindane on the soil microfauna of a spruce forest: A field study using a completely randomized block design. *Biol. Fert. Soils,* **7**, 225–231.

Design of a Sampling Regime for Field Studies to Investigate the Effects of Agrochemicals on Earthworms

J.M. COULSON, E.C. McINDOE, P.F. CHAPMAN AND P.J. EDWARDS

ICI Agrochemicals, Jealott's Hill, Bracknell, Berkshire RG12 6EY, UK

Introduction

It may be necessary after conducting laboratory work to go on to investigate the effects of agrochemicals on earthworms in the field. ICI Agrochemicals has carried out field testing on earthworms since the 1960's and details of the grassland plot trials were published by Edwards and Brown in 1982. In addition, trials have been conducted in agricultural crops. This paper draws on the data generated to investigate the size of effects that may be detected by sampling schemes with different numbers of replicates and subsamples. In addition, data for different groups of earthworms are presented.

Experimental design

In such studies, the typical experimental design is a randomised block consisting of several treatment groups and a control, each replicated several times. Within each plot, a number of subsamples are usually taken to obtain as precise an estimate as possible of the mean response for that plot. An estimate of experimental error is then found from analysis of variance. This is then used to construct t-tests to contrast the mean effect of each treatment against the mean of the control.

If any of the treatments has a biologically significant effect on earthworm numbers, it is important that we maximise the chances of detecting this effect. We decide that an effect is real when we reject the null hypothesis of no treatment effect, ie. when we observe a statistically significant result. We therefore choose the number of replicates and the number of subsamples in order to produce a test which will maximise the probability of observing a statistically significant result when the effect of the treatment is biologically important.

Variation in earthworm numbers in field trials has been investigated by comparing the within plot variance with the between plot variance. In the table below the ratio of between/within plot variance is presented for four sites for the following groups of earthworms: *Lumbricus* spp., *Allolobophora* spp., total adults, total immatures and total earthworms.

Table 1. Between Plot/Within Plot Variance for Groups of Earthworms on Four Trial Sites

Group/Site	1	2	3	4
Lumbricus	0·37	1·79	1·00	0·33
Allolobophora	2·09	0·80	0·69	2·80
Adults	0·75	0·50	0·32	0·15
Immatures	1·34	1·17	0·78	6·26
Total	1·00	0·76	1·17	2·78

It can be seen from these data that the ratios for all the groups are similar and that a typical value for the between/within plot ratio is about 1. Hence, this value has been used when comparing sampling schemes below.

The following figure shows the probability of rejecting the null hypothesis v percentage difference in earthworms numbers for nine sampling schemes.

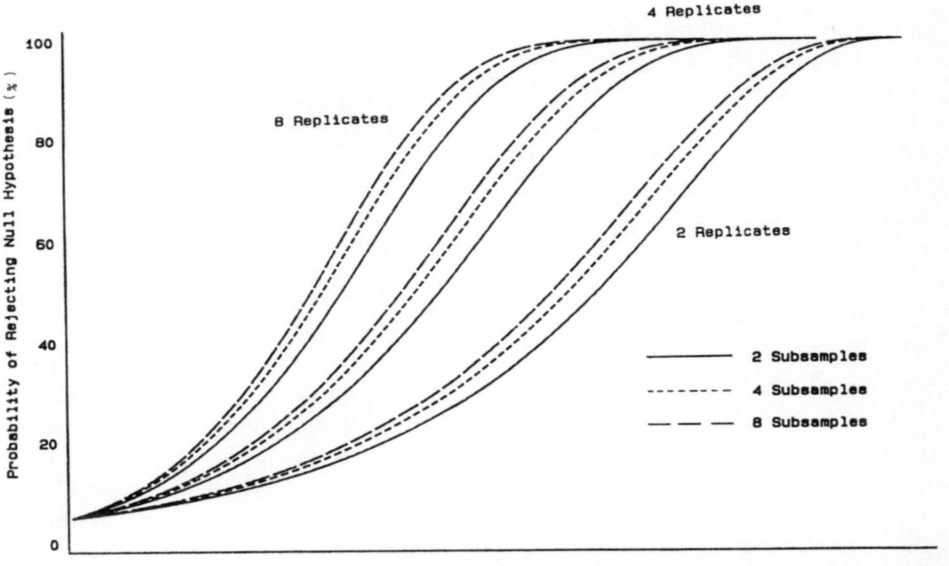

Figure 1. Probability of rejecting null hypothesis v % reduction in earthworm numbers for 9 different sampling schemes between-plot variance = within-plot variance = 0·025 (log (no.))^2

The figure shows that more benefit is derived from greater amounts of replication than subsampling. However, the amount of resource required for additional replication is greater than that for additional subsampling. For field earthworm trials this difference is small, probably being only double the resource.

Conclusions

1. The ratio of between plot and within plot variance is approximately 1 for a typical earthworm field trial. This applies for the biological groups: *Lumbricus, Allolobophora*, adults and immatures as well as total earthworms.
2. Given typical levels of variability, greater replication is of more benefit than greater subsampling. However, replication requires approximately twice as much resource as subsampling.
3. If the size of an effect of biological importance can be decided, then by statistically comparing sampling regimes, a trial design can be produced that can detect that effect as statistically significant.

References

EDWARDS, P.J. AND BROWN, S.M. (1982). Use of grassland plots to study the effects of pesticides on earthworms. *Pedobiologia,* **24**, 145–150.

Assessing Soil Metal Pollution with Earthworms: Indices Derived from Regression Analyses

J.E. MORGAN, A.J. MORGAN AND N. CORP

School of Pure and Applied Biology, University of Wales College of Cardiff, P.O. Box 915, Cardiff CF1 3TL, UK

Introduction

An important objective of ecotoxicology is to evaluate the degree of pollution within a site or sites. In the case of metal-contaminated soils, direct chemical methods provide data concerning the 'total' and 'soluble-fraction' concentrations. But, because of the plethora of interactive edaphic factors, they cannot easily provide accurate estimates of the availability of metals to specific biotic components. An alternative approach is to use accumulated concentrations of metals in the tissues of suitable monitoring organisms as integrative estimates of the relative degrees of metal pollution. This conceptually attractive idea is, however, fraught with practical difficulties. Firstly, even closely related species accumulate significantly different metal concentrations (Greville & Morgan, 1989; Hopkin *et al.*, 1985; Morgan & Morgan, 1991). Secondly, age, size, season and diet are factors that influence metal accumulation by a given species.

Boyden (1977) suggested using a comparison of the slopes of regression lines from log-log plots of metal content versus dry body weight in marine molluscs for comparing pollution at different sites. The principle of the method is illustrated in Fig. 1A. It takes into account variability due to species and body weight differences; it also provides a pollution index independent of seasonal and environmental fluxes in metal concentrations. Simkiss & Taylor (1981) proposed an extension of this approach, where the ratios of the intercept values for two metal ions could be used to compare and predict the degree of metal pollution. However, significant differences can easily be overlooked if both components of the ratio change in similar ways. A third approach is to plot regression slopes against intercept values to provide size-independent comparisons of relative metal bioavailabilities (Burton *et al.*, 1983; Morgan, 1987).

Earthworms are efficient accumulators of soil metals, the proportion of the

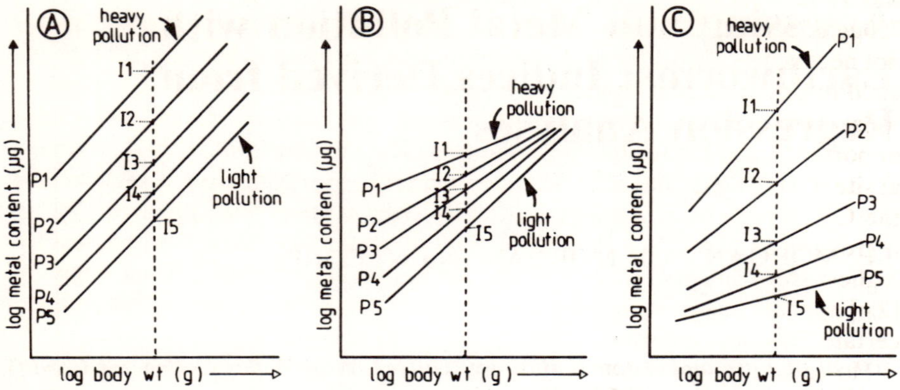

Figure 1. Hypothetical log-log plots of metal *content* against dry body weight from sites with different degrees of pollution (P1–P5). The intercept values (I1–I5) are normalized to a convenient weight (near the geometric mean) to remove extrapolation errors. A = slopes are roughly parallel, intercepts differ ('Boyden's model'). B = as degree of pollution increases, the intercepts increase but slopes decrease, implying a measure of metal regulation. C = as degree of pollution increases, both the intercepts and slopes increase, implying that 'tolerance' is expressed as enhanced accumulation capacity.

'total' soil metal available to them depending on a number of edaphic factors (Ma, 1982; Morgan & Morgan, 1988; Kiewiet & Ma, 1991). In our experience the regression slopes (log transformed metal-content versus body weight) for different earthworm populations are not parallel, and differ from metal to metal (compare: Figs. 1A, 1B and 1C). Furthermore, heavy natural selection pressures at specific contaminated sites could be expressed as physiological adaptations that confound simple predictive schemes (Klerks & Levinton, 1989). For these reasons, this paper explores the biomonitoring possibilities of slope-intercept plots, and examines the circumstances under which they can be effectively applied.

Materials and methods

Clitellate earthworms (*Dendrodrilus rubidus* and *Lumbricus rubellus*) were sampled from several disused Pb/Zn mine sites in Wales and Shropshire, England, and from one uncontaminated site. Because the two species were collected from these notoriously heterogeneous sites on different occasions, the sites in the *D. rubidus* plots are referred to as numbers, and in the *L. rubellus* plots as letters.

Whole-worm and soil digests (conc. HNO_3) were analysed for Zn and Cd by atomic absorption spectrophotometry.

Results and discussion

The Zn and Cd concentrations in *D. rubidus* and *L. rubellus* (Fig. 2) were significantly and positively correlated with 'total' soil metal concentrations.

In general, the Zn concentration factor (C.F. = worm [Zn] ÷ soil [Zn]) was ›1 in soils with a [Zn] ‹ 750μg/g, and a C.F. ‹1 for soils with a [Zn] › 750μg/g; the Cd C.F. was usually ›1. These 'simple' linear relationships are not necessarily useful for pollution monitoring purposes, because they do not yield information that cannot more easily be derived by the direct analysis of soils. However, the presence of 'outlier' populations reveal potentially important local bioavailability conditions. For example, [Cd] in *L. rubellus* at site 'f' (Fig. 2C), and in *D. rubidus* at site '5' (Fig. 2D), are unusual because the C.F.'s are ‹1 i.e. the worm [Cd] was much lower than predicted by the regression analyses. Scrutiny of the data indicated that these sites possess the highest soil [Zn] in each case (Figs. 2A & 2B, respectively). Thus, high soil [Zn] appears to suppress the accumulation of Cd, a metal with which it shares certain chemical affinities.

However, tissue concentration values are subject to several sources of variability (Boyden, 1977). Plots of the regression slopes against intercept values, derived by plotting metal content against body weight (log-log

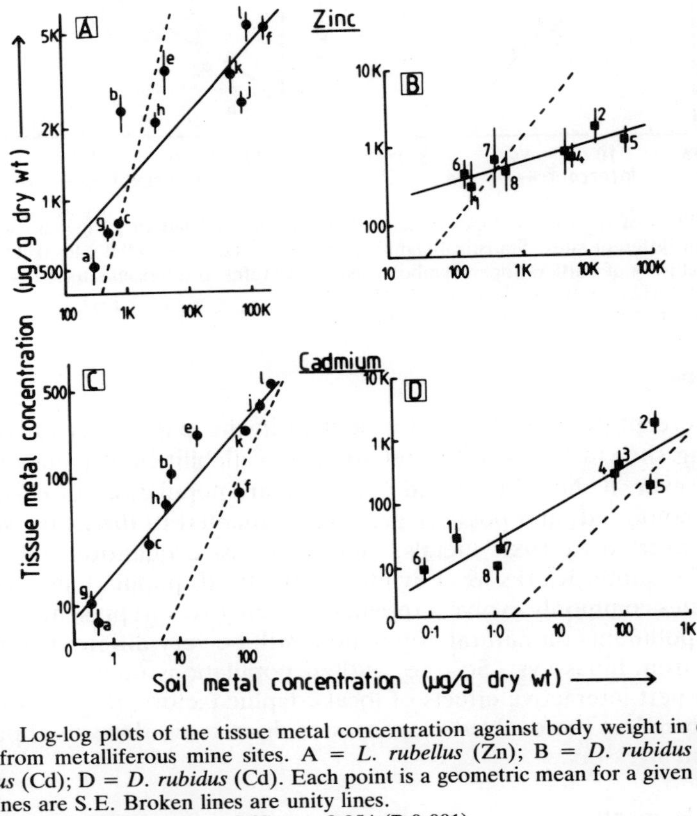

Figure 2. Log-log plots of the tissue metal concentration against body weight in earthworms sampled from metalliferous mine sites. A = *L. rubellus* (Zn); B = *D. rubidus* (Zn); C = *L. rubellus* (Cd); D = *D. rubidus* (Cd). Each point is a geometric mean for a given population, vertical lines are S.E. Broken lines are unity lines.

A: $\log_{10}\mathbf{y} = 0{\cdot}184 \log_{10}\mathbf{x} + \log_{10} 2{\cdot}27$; r = 0·854 (P‹0·001)
B: $\log_{10}\mathbf{y} = 0{\cdot}219 \log_{10}\mathbf{x} + \log_{10} 2{\cdot}11$; r = 0·739 (P‹0·001)
C: $\log_{10}\mathbf{y} = 0{\cdot}307 \log_{10}\mathbf{x} + \log_{10} 1{\cdot}12$; r = 0·860 (P‹0·001)
D: $\log_{10}\mathbf{y} = 0{\cdot}582 \log_{10}\mathbf{x} + \log_{10} 1{\cdot}29$; r = 0·910 (P‹0·001)

transformed, and normalized to 25 mg) for *D. rubidus*, indicated: (a) that tissue [Zn] decreased with respect to body weight with increasing soil [Zn] (see Fig. 3A, *c.f.* Fig. 1B); (b) tissue [Cd] increased with respect to body weight with increasing soil [Cd] (see Fig. 3B; *c.f.* Fig. 1C). These weight normalized plots revealed certain specific outlier populations, that were not identified in the log-log transformed plots of tissue metal concentrations against soil metal concentrations. First, the Zn at site '2' appears to be either more 'bioavailable', or less well regulated, by the worms occupying the site (Fig. 3A); interestingly, site '2' possessed the highest degree of Cd contamination (Fig. 3B). Second, the Cd at site '3' is less bioavailable, or is regulated by earthworms belonging to this specific population (Fig. 3B), although no explanation is immediately obvious.

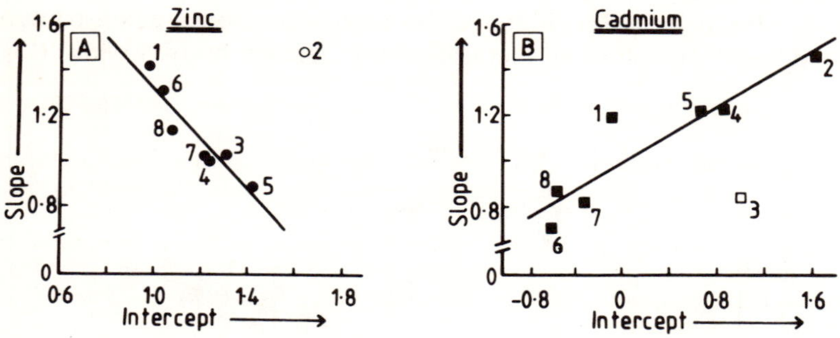

Figure 3. Plots of regression slopes versus intercepts (normalized to 25 mg dry wt.) for *D. rubidus* from different sites. Statistical significance was tested by Kendall's Rank Correlation, after the exclusion of outliers (open symbols). Numbers refer to different sites.
A (Zn): $y = 2{\cdot}412 - 1{\cdot}104\,x$; $rk = -0{\cdot}810$ ($P<0{\cdot}01$)
B (Cd): $y = 1{\cdot}004 + 0{\cdot}294\,x$; $rk = 0{\cdot}905$ ($P<0{\cdot}005$)

Conclusions

Slope: intercept plots of the kind described in this paper may prove useful for ranking sites in terms of the relative bioavailabilities of metals at those sites. However, it should be noted that these are populations of earthworms that have colonized, and possibly genetically adapted to the heavy selection pressures exerted by toxic metals, and/or excessive quantities of essential metals. To quote Klerks & Levinton (1989): 'if populations inhabiting polluted sites commonly evolve a resistance to the toxicant present, long-term effects of pollutants on natural populations will be very different from those predicted from bioassays'. So, the 'outlier' populations may not exclusively reflect the nett interactive effects of local edaphic factors; they may to some extent also reflect the evolved responses of discrete earthworm populations to local soil stressors.

Acknowledgements

We thank Alison Richards for typing the manuscript, and Vyv Williams for assistance with the figures.

References

BOYDEN, C.R. (1977). Effect of size upon metal content of shellfish. *J. mar. biol. Ass. U.K.* **57**, 675–714.

BURTON, K.W., MORGAN, E. AND WILLIAMS, A.T. (1983). Trace heavy metals in *Fucus* and *P. vulgata* along the Glamorgan Heritage Coast, South Wales. *Water, Air and Soil Pollut.* **19**, 377–388.

GREVILLE, R.W. AND MORGAN, A.J. (1989). Concentrations of metals (Cu, Pb, Cd, Zn and Ca) in six species of British terrestrial gastropod living near a disused Pb/Zn mine. *J. Moll. Stud.* **56**, 355–362.

HOPKIN, S.P., MARTIN, M.H. AND MOSS, S.J. (1985). Heavy metals in isopods from the supra-littoral zone on the southern shore of the Severn Estuary, U.K. *Environ. Pollut.* **9**, 239–254.

KIEWIET, A.T. AND MA, W.-C. (1991). Effect of pH and calcium on lead and cadmium uptake by earthworms in water. *Ecotox. Environ. Safety* **21**, 32–37.

KLERKS, P.L. AND LEVINTON, J.S. (1989). Rapid evolution of metal resistance in a benthic oligochaete inhabiting a metal-polluted site. *Biol. Bull.* **176**, 135–141.

MA, W.-C. (1982). The influence of soil properties and worm-related factors on the concentration of heavy metals in earthworms. *Pedobiologia* **24**, 109–119.

MORGAN, J.E. (1987). Earthworms as biological monitors of cadmium and zinc in highly contaminated metalliferous soils. *J. South East England Soil Discussion Group* **5**, 24–37.

MORGAN, J.E. AND MORGAN, A.J. (1988). Earthworms as biological monitors of cadmium, copper, lead and zinc in metalliferous soils. *Environ. Pollut.* **54**, 123–138.

MORGAN, J.E. AND MORGAN. A.J. (1991). Differences in the accumulated metal concentrations in two epigeic earthworm species (*Lumbricus rubellus and Dendrodrilus rubidus*) living in contaminated soils. *Bull. Environ. Contam. Toxicol,* **47**, 296–301.

SIMKISS, K. AND TAYLOR, M. (1981). Cellular mechanisms of metal ion detoxification and some new indices of pollution. *Aquatic Toxicol.* **1**, 279–290.

Effects of Composted Sewage Sludge on the Earthworm *Lumbricus terrestris*

C.L. FORDHAM[1] AND C.G. WILBER[2]

[1] *Terra Technologies, Golden, Colorado, USA*
[2] *Colorado State University, Fort Collins, Colorado, USA*

Abstract

The purpose of this study was to investigate bioavailability of cadmium and lead in sewage sludge-amended alkaline soils, as well as to investigate toxicity to earthworms of composted sewage sludge. Two alkaline soil types were selected for study. Composted sewage sludge was obtained from the Denver Metropolitan Sewage Treatment Plant. Earthworms (*Lumbricus terrestris*) were exposed to different ratios of composted sewage sludge and soil, and to clean soil only as controls. Earthworms were observed for a treatment period of up to 20 days, then killed and analyzed for metal content. A dose-response relationship was observed between mortality and increasing sludge content of soil. Mean tissue cadmium and lead concentrations were not significantly related to soil type or treatment level.

Introduction

Land application of sewage sludge has been considered as a method of disposal. Composting sewage sludge with wood chips helps facilitate land disposal. There are, however, unanswered ecological questions pertaining to the practice of land disposal of sewage sludge. Sludge from metropolitan areas typically contains industrial contaminants such as heavy metals (Brown *et al.*, 1983). The bulk of all trace metal accumulation resulting from sludge application is in the upper portion (15 cm) of the soil profile (Chang *et al.*, 1979), a zone where soil fauna are active and may be exposed.

The purpose of this study was to investigate bioavailability of cadmium and lead in sewage sludge-amended alkaline soils, as well as to investigate toxicity to earthworms of composted sewage sludge. Cadmium was investigated due to its toxicity to earthworms and higher trophic level organisms, and its potential to bioaccumulate from earthworms to other species. Lead was investigated due to its ubiquity in sewage sludge and soil. The toxicity of sewage sludge was investigated in order to evaluate potential hazard to the soil faunal community and resulting impacts to the terrestrial food web.

Methods

Two alkaline soil types were selected for study – Ascalon and Weld. Sewage sludge was obtained from the Metropolitan Sewage Treatment Plant in Denver, Colorado. Soils and sludge were oven-dried and then sieved to produce a uniform particle size. Soils and sludge were analyzed for both extractable and bioavailable cadmium and lead, and pH. Bioavailable metals were determined by the NH_4HCO_3-DTPA method (Soltanpour and Workman, 1979; Havlin and Soltenpour, 1981). Flame atomic absorption spectrophotometry was used for other metal analyses.

Earthworms were purchased from Carolina Biological Supply, North Carolina. Earthworms were maintained at temperatures of 16 to 18°C on clean, moist filter paper for 48 hours prior to being placed in treatment containers. Treatment containers measured 22·9 by 17·8 by 7·6 cm.

Three treatments and a control were used for each soil type in the study. Treatments were designed to simulate, double, and triple the agricultural application rate of 25 to 30 tons of sludge per acre. The treatments were 17, 34, and 51% sewage sludge:soil. Dried sludge was thoroughly mixed with dried soil on a weight to weight basis before adding earthworms to the containers. Twenty-five earthworms were randomly selected and placed in each of the control and two lower treatments, and 10 earthworms were placed in each of the highest treatments.

Earthworms were observed for mortality and sampled at 3 day intervals for one week and then 5 day intervals for the duration of the study (20 days). Earthworms were rinsed with deionized water to remove extraneous material, then killed by freezing. Gut contents were flushed by dissecting the earthworm lengthwise and rinsing gently with deionized water. Earthworms were oven dried, weighed, and placed in individual test tubes. A nitric-perchloric acid digestion was performed, and samples were analyzed on an atomic absorption spectrophotometer. Earthworms were not pooled, but analyzed individually.

Statistical methods included testing mortality data with a chi square test of significance comparing *m* cohorts for a dichotomous outcome (Fleiss, 1981). The level of probability was 0·05. Tissue metals were tested with MANOVA, Student-Newman-Keuls procedure, and Duncan's multiple range test at a level of significance of 0·05–0·10.

Results

Dead and moribund earthworms were removed from the test containers on each sampling date. Moribund and dead earthworms were observed to have ulcerations over the length of the body; lesions were more prevalent on the anterior end. The clitellum was enlarged and in some cases ruptured. Sexual organs were often enlarged and eviscerated, and a thick coat of mucus covered the body. Body fluids extruded from the earthworm were apparent on the surrounding soil. Earthworms were limp and lying on the soil surface, and there was no response to gentle prodding. Mortality differed with soil type, and increased with dose.

The highest treatment level (51% sludge) produced immediate response in earthworms, which began to writhe on contact with the material and refused to bury. All earthworms in the higher treatment levels were dead within the first six days of the study. By the end of the study, earthworms in the lower treatment levels were also lying on the soil surface or were dead. No control earthworms died within the time of the experiment. Total mortality in the lowest treatments (17% sludge) differed significantly between soil types (Ascalon > Weld).

Cadmium concentrations tended to be higher only in earthworms exposed to the lowest sludge level (17%) as compared to controls, and higher in earthworms from Weld soils as compared to Ascalon (Figure 1). Lead concentrations in the Ascalon control earthworms tended to be higher than in the earthworms from the two lowest treatment groups; but earthworms exposed to sludge:Weld soil had higher or similar tissue lead concentrations as controls (Figure 2). Differences in tissue concentrations were not statistically significant for soil type or treatment level for either cadmium or lead. This may relate to overall high variability and possibly the rapid mortality observed in the 34 and 51% treatment groups. Tissue concentrations of cadmium were observed to correlate negatively with earthworm body weight ($p < 0{\cdot}1$), although lead data did not approach significance.

Bioaccumulation factors for lead were <1 in earthworms from Ascalon soil and <2 in earthworms from Weld soil. The cadmium concentration factors were 35 for earthworms from Ascalon soil and 76 for earthworms from Weld soil for the lowest exposed group (17%) on day 20. Cadmium bioaccumulation factors were elevated with respect to controls, which were 21 for Ascalon and 39 for Weld. Chemical and physical characteristics of sludge and soils are given in Table 1.

Discussion

The composted sewage sludge was obviously toxic to earthworms. Mortality was significantly higher in the lowest treatment of the Ascalon compared to the Weld soil mixture, an effect possibly related to the lower pH of Ascalon

Table 1. Physical characteristics of Ascalon and Weld soils, and composted sewage sludge. The values are given for pH, organic matter percent, bioavailable and total extractable metals.

Material	pH	Organic Matter (%)	Bioavailable Metals (AB-DTPA)	Extractable Metals
Sludge	4·7	80 n = 3	5·9 ppm Cd	19·3 ± 1·18 ppm Cd n = 6
			15·5 ppm Pb	294·5 ± 19·45 ppm Pb
Ascalon	7·0	2·94 n = 5	0·08 ppm Cd	2·6 ± 0·29 ppm Cd n = 3
			13·5 ppm Pb	39·0 ± 0·99 ppm Pb
Weld	7·6	1·20 n = 5	0·06 ppm Cd	3·7 ± 0·61 ppm Cd n = 3
			2·9 ppm Pb	33·3 ± 0·99 ppm Pb

AGRICULTURAL ZOOLOGY REVIEWS VOLUME SIX

Editor: K. EVANS.

Expected publication date September 1993; ISBN 0 946707 51 0;
ISSN 0269-0543; 234 x 156mm; approx 400pp; illustrated; index; hard cover;
Price £70.00

Provisional contents:

- Regulatory Control of Invertebrate Crop Pests. PAUL BARTLETT, JOHN COTTER (UK)
- Pest Control Problems on Potatoes Grown in Andean Regions. J. FRANCO, M. CANTO (Peru)
- Targetting Techniques in Pesticide Application. M.G. FORD (UK)
- Interactions Between Plants and Insect Parasitoids. S. UDAYAGIRI (USA)
- Entomopathogenic Nematodes. R. GEORGIS (USA)
- The Screw-worm Problem. J.P. SPRADBERY (Australia)
- Groundnut Entomology: retrospective and prospective view. R.T. GAHUKAR (India)
- The Agricultural Importance of Slugs. I.F. HENDERSON, A.W. FERGUSON (UK)
- Plant Resistance to Insects. G.R. MANGLITZ, S. DAVIDSON (USA)
- The Dependence of Major Crops on Bees for Pollination. T. H. WILLIAMS, D.G. ALFORD (UK)
- Colorado Potato Beetles. R.F.W. SHRODER (USA)

- ORDER FORM -

Return to: Intercept Limited, P.O. Box 716,
Andover, Hampshire, SP10 1YG, England, U.K.
Tel: 0264 334748 Fax: 0264 334058

Please send me this title....

Quantity

Postage: £2.50 - UK
£3.00 - overseas

❑ by cheque (attached herewith)

❑ please charge my credit card:

❑ Mastercard ❑ Visa ❑ Amex

Card No. ____________________
Expiry Date: ____________________
Signature: ____________________

Proforma Invoice:
Please send me a proforma invoice including packing and postage by:

❑ Surface Mail ❑ Air Mail

Other terms:

❑ Bill me: enter Purchase Order No.

____

Please tick appropriate boxes.

soil. Bioavailable metals were higher in this soil type as well. Metal mobility and thus bioavailability increase inversely to soil pH; thus, toxicity may also increase. Differences in mortality between soil types in the two higher dose levels were not significant. The soil concentrations of cadmium and lead were probably not the cause of the observed mortality, however. A six-week LC_{50} for cadmium ranges from 150 to 1000 ppm, and a six-week LC_{50} for lead is in excess of 3000 ppm (Ma, 1982). Observed concentrations of these two heavy metals were much less than these reported toxic concentrations. Other metals were not analyzed.

Earthworms from the higher treatment levels tended to have concentrations of lead similar or lower than earthworms from the lowest treatment, or the controls. This may be a result of the toxicity response observed in earthworms from the two higher treatments, where coelomic fluid was observed on the soil surface. Part of the earthworm body burden of metals is contained within the coelomic fluid (Andersen and Laursen, 1982), and loss of this fluid can thus reduce the observed body burden. Cadmium results in treatment compared to controls possibly were confounded by the observation that tissue cadmium decreased with increasing body weight in earthworms from both soil types. Cadmium is bound to a metallothionien binding protein and stored in the gut wall, whereas lead is stored in brown bodies and shed through the dorsal pores (Andersen and Laursen, 1982). Increased loss of coelomic fluid could therefore increase the excretion of lead to a greater extent than that of cadmium. Cadmium concentrations were elevated at Day 20 as indicated by increased bioaccumulation factors with respect to the controls. There was no indication that excessive bioaccumulation of lead occurred.

Since bioavailable metals were higher in Ascalon soil as opposed to Weld, bioaccumulation factors would have also been expected to be higher. The fact that they were lower suggests that the higher toxic response in earthworms from Ascalon soil could have reduced bioaccumulation, or that results were possibly confounded by body size. Standardizing body size among the test animals should be done to eliminate this source of variability. Samples should consist of two or more pooled individuals, or sample size should be doubled to obtain stronger statistical analysis.

The lowest ratio of composted sewage sludge to soil was intended to simulate field application rates. Based on the observations of mortality it is possible that earthworms exposed to composted sewage sludge at these rates in the field may be adversely affected. Organisms that depend on earthworms as a food source may also be adversely impacted as a result of loss of prey.

Earthworms are valuable tools for investigating ecological effects of contaminants. They are sensitive to many chemicals, are readily maintained test organisms, and can be used to investigate potential food web effects due to bioaccumulative contaminants. Variables that need to be accounted for when using earthworms to assess ecological risk include soil type and soil parameters, species and population specific parameters (i.e., tolerance), organism size, ambient field parameters (i.e., temperature, moisture), bioassay methodology, and waste characterization.

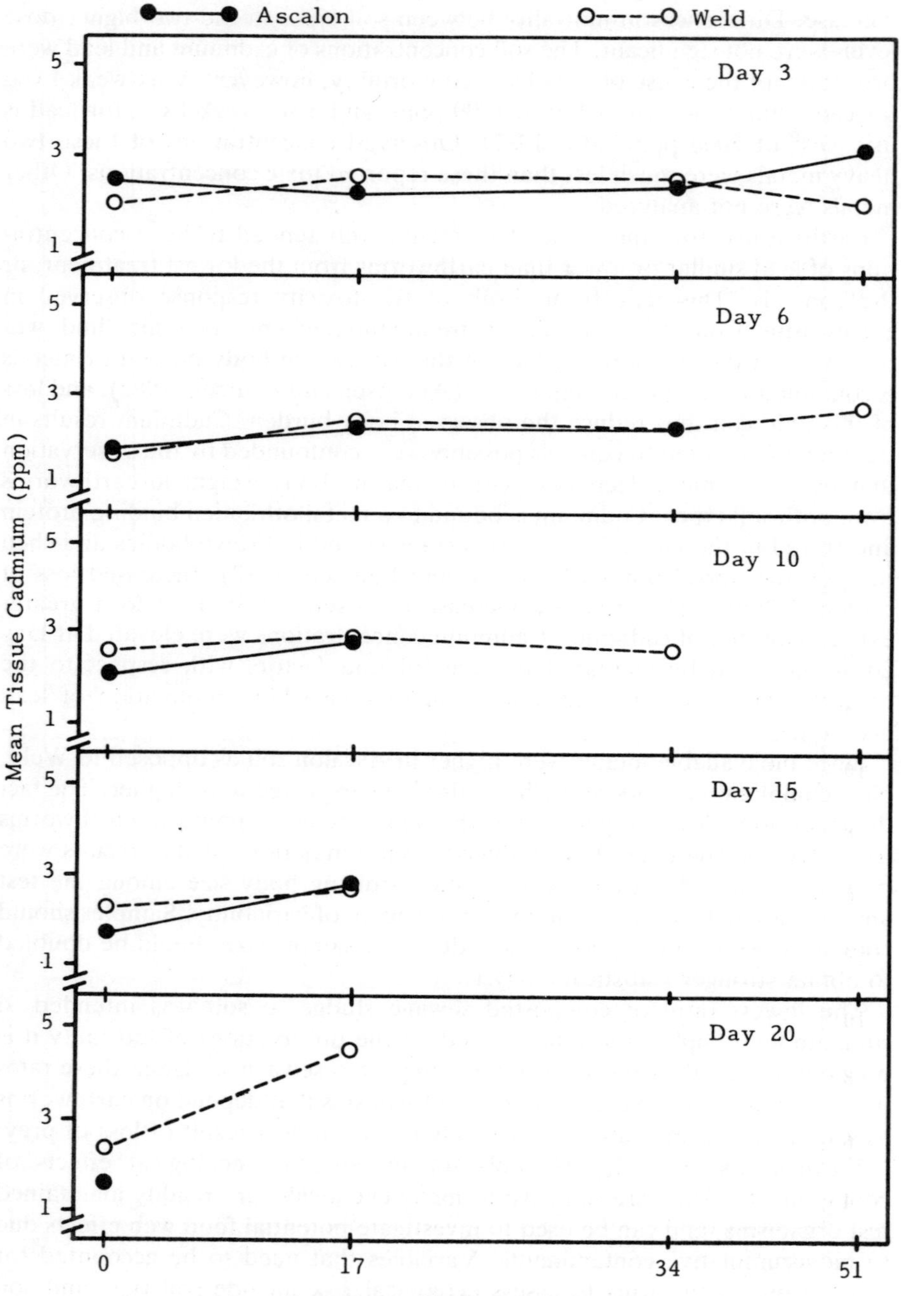

Figure 1. Tissue Cadmium Concentrations by Soil Type, Time, and Log Dose

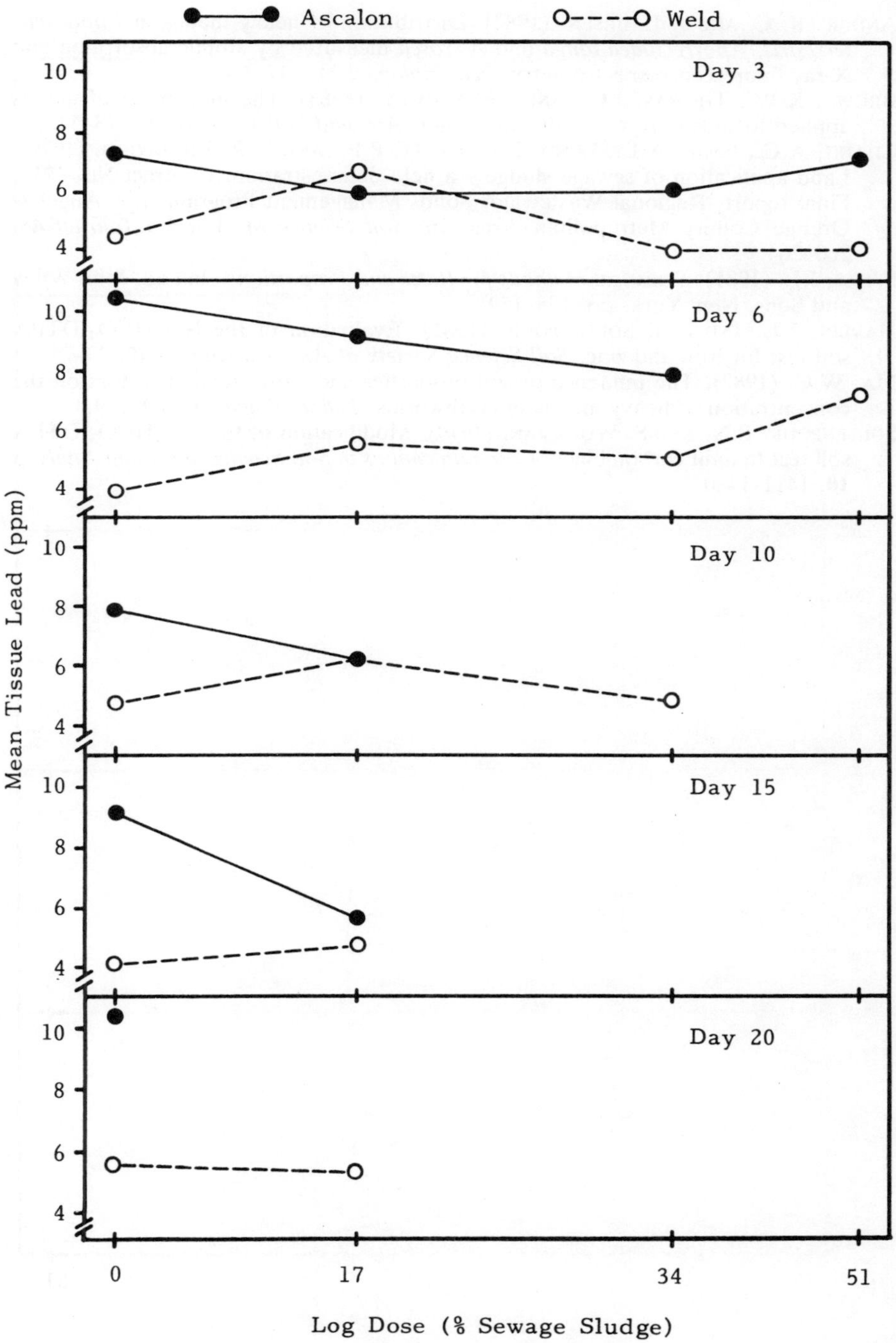

Figure 2. Tissue Lead Concentrations by Soil Type, Time, and Log Dose

References

ANDERSEN, C. AND J. LAURSEN. (1982). Distribution of heavy metals in *Lumbricus terrestris, Aporrectodea longa* and *A. rosea* measured by atomic absorption and X-ray fluorescence spectrometry. *Pedobiologica* **24**, 347–356.

BROWN, K.W., THOMAS, J.C., AND J.F. SLOWEY. (1983). The movement of metals applied to soils in sewage effluent. *Water, Air, and Soil Pollution* **19**, 43–54.

CHANG, A.C., PAGE, A.L., LUND, L.J., PRATT, P.F., AND G.R. BRADFORD. (1979). Land application of sewage sludge – a field demonstration. Contract No. 2218. Final report. Regional Wastewater Solids Management Program, Los Angeles/ Orange County Metropolitan Area. In: *Soil Science of America Journal* **46**, 260–264.

FLEISS, J.L. (1981). *Statistical Methods for Rates and Proportions*. 2nd ed. John Wiley and Sons, New York. pp. 138–159.

HAVLIN, J.L. AND P.N. SOLTENPOUR. (1981). Evaluation of the NH_4HCO_3-DTPA soil test for iron and zinc. *Soil Science Society of America Journal* **45**, 70–75.

MA, W.C. (1982). The influence of soil properties and worm-related factors on the concentration of heavy metals in earthworms. *Pedobiologia* **24**, 109–119.

SOLTENPOUR, P.N. AND S. WORKMAN. (1979). Modification of the NH_4HCO_3-DTPA soil test to omit carbon black. *Communications in Soil Science and Plant Analysis* **10**, 1411–1420.

Recommendations

Recommendations of an International Workshop on Ecotoxicology of Earthworms

P.W. GREIG-SMITH

Sheffield, UK, April 1991

Introduction

A specialist Workshop was held on 10–12 April, 1991 at the University of Sheffield, UK, in association with the first European conference of the Society of Environmental Toxicology and Chemistry (SETAC). There were 60 participants from 13 countries, who contributed to an active programme of invited talks, posters and discussion sessions. The organising committee were: H. Becker (BBA, Germany) (Chairman), C. Callahan (Environmental Protection Agency, USA), C.A. Edwards (Ohio State University, USA), P.J. Edwards (ICI Agrochemicals, UK), P.W. Greig-Smith (MAFF, UK), and F. Heimbach (Bayer AG, Germany).

The aims of the Workshop were to assemble an international group of experts in earthworm ecotoxicology, and those with an interest in the implications of such studies, in order to address three objectives. First, to review the state of the art in this field, covering knowledge of techniques, test protocols, principles of interpretation, and practical experience. Second, to identify gaps in knowledge, on which new research and development work should be initiated. Third, to clarify issues on which there is disagreement or debate, and attempt to harmonize approaches to test methodology.

Although the interests of participants were diverse, a common unifying theme was the desire to construct a framework for a stepwise, flexible sequence of testing for toxic effects. This should involve successive tiers of testing, each invoked only if necessary, to progressively clarify whether a chemical is substantially harmful to earthworms. Tests should allow an assessor to classify effects from ‘high risk’ to ‘no risk’, in order to make decisions about approval of products. Generally, only if initial results indicate a potential for substantial effects in the field would more detailed, higher-tier studies be required.

Figure 1 illustrates the principles of the successive stages of assessment, which relate to the four main themes of the Workshop, on Laboratory Tests, Field Tests, Estimation of Exposure, and Risk Assessment. These

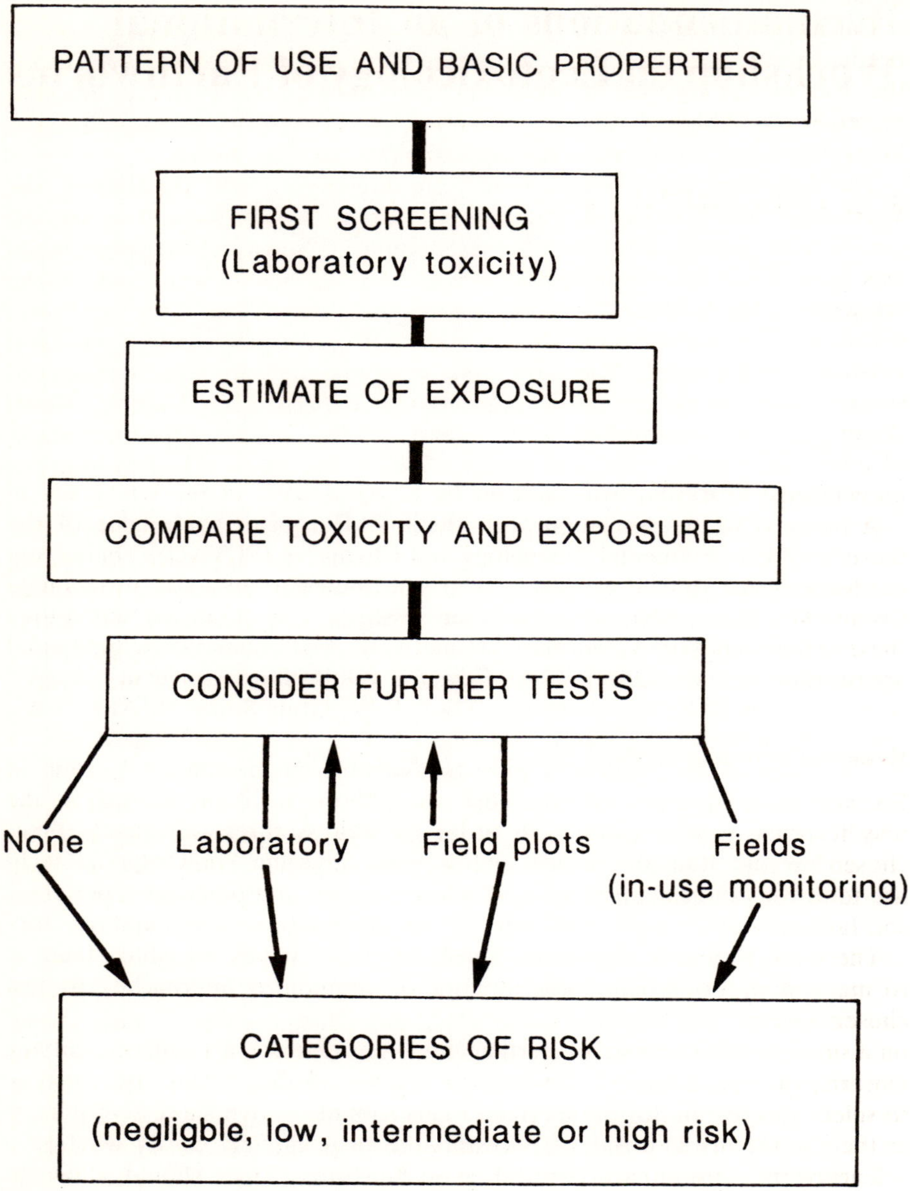

Figure 1. Outline of the stepwise sequence of assessment for the effects of contaminants on earthworms

recommendations are structured in the same way, considering in turn the conclusions from discussions about these aspects. In addition, there was a preliminary discussion that tackled the question of why tests need to be conducted on earthworms, plus two sessions dealing with specific topics – the significance of pesticide residues in earthworms, and the use of earthworms in monitoring studies.

Earthworm toxicology is particularly relevant for the registration of pesticides and other chemicals. There are also other implications, such as for the use of earthworms in assessing contaminated sites and monitoring remediation. Although the emphasis was placed firmly on pesticides, the Workshop also considered other contaminants, such as metals.

The following sections are based on discussions held throughout the Workshop, which included lengthy open forum debates as well as specific questions at the end of the formal presentations. A record of the points made was kept by P.W. Greig-Smith and C. Aldridge, and was used as the foundation for a draft of these recommendations. The draft was then distributed for comment to all the workshop participants before the final version was prepared. Therefore, the recommendations represent a consensus view among the many international experts from Europe, North America, New Zealand and South Africa, on the principles for assessment of effects on earthworms, and on what action should be taken to improve knowledge and regulatory approaches to these issues.

A majority of the participants were from Europe, and it is in Europe that there is greatest current concern to rationalise the approach to testing earthworms for regulatory purposes, because of the impending European Community harmonization of pesticide registration procedures. However, these recommendations have been written to provide more general guidance, appropriate to earthworm and contaminant issues in all parts of the world.

Reasons for testing earthworms

For any assessment of environmental risk, it is important to be clear about why tests are being carried out. In the case of earthworms, they may be chosen because of their beneficial role in promoting soil fertility, because they are food for other animals, or as convenient general representatives of the soil fauna.

There are two basic approaches to the selection of test organisms. One is to make a statistical conclusion from a suitable range of often arbitrarily chosen species; one example is the 'HC5' procedure, by which toxicity tests on a small number of species are used to decide on the level of safety for the majority of the soil fauna (Van Straalen and Denneman, 1989). The other is to select specific 'indicator' species on the basis of particular features, such as their sensitivity to poisoning, likelihood of exposure, and ease of study.

Earthworms are particularly important because of their relevance to the structure of soil and to nutrient cycling. This is a consequence of their relatively large size, their behaviour, and their high biomass in some areas. For certain types of contaminants, they are also more sensitive than many other animals, and therefore their protection may provide a margin of safety for other fauna.

A different advantage of earthworms lies in their suitability for laboratory testing, being convenient for handling, suitable for captive breeding, and readily available. Of course, these features vary among earthworm species so that a separate issue of choosing species for testing arises (see later).

In the selection of soil organisms for testing, it is important to define whether concern is for the *structure* of the soil community (e.g. species diversity, biomass, trophic relationships) or for its *function* (e.g. decomposition of organic material, mineralization of nutrients, synthesis of humus). Generally, functions can be maintained by protecting structure, but preservation of functions does not necessarily mean that community structure is unchanged. Earthworms have well-known functional properties, and protection of earthworms is one way to support these processes.

RECOMMENDATION 1

Ensure that the purpose of any proposed ecotoxicological assessment involving earthworms is clearly defined, so as to allow the relevance of earthworm tests to be properly judged.

RECOMMENDATION 2

Encourage the development and review of statistical procedures for using the results of tests on soil organisms, in order to assess their validity compared to case by case evaluation of selected indicator species, including earthworms.

Effects in the laboratory

There are a wide range of techniques and approaches available for the study of effects of chemicals on earthworms under controlled, experimental conditions in the laboratory. This has led to the production of several protocols, advocated as suitable for adoption as standard test methods. Some are well-established, widely-used methods that have changed little since they were first proposed (e.g. the OECD Method No 207; OECD, 1984). There are other protocols, which are relatively new, or still under development (e.g. reproduction tests; Van Gestel, 1989; Kokta, 1992). Some of these, for example the Daniel funnel test (Bieri, 1989), protocols using Tubifex (Ammon, 1985) or enchytraeid species and electrophysiological methods (Drewes, 1992) provide novel means of testing for effects, but are not widely used.

The advantages and disadvantages of these methods vary greatly, in the degree of standardization that is possible, their convenience and practicality, the accuracy and precision of their results, and relevance to effects in the field.

Which methods are most appropriate depends on the purpose of the tests to be carried out. A test may serve as a preliminary screen of chemicals to identify those that require further investigation; it may help to model the fate, exposure and effects of a chemical in the field; or it may be simply a research tool for the study of toxic mechanisms.

The aim of a screening test is to obtain a reliable indication of the likely safety for earthworm populations of a chemical or product when used under field conditions. The greatest priority is to identify chemicals which do not have significant effects, while avoiding false negatives. Those that do show effects at relevant concentrations, or for which results are uncertain, should then be subjected to further study.

Tests to model effects in the field are different. They require estimates of known accuracy; they should simulate actual conditions of use as closely as possible; and they should involve relevant species, or at least allow reliable extrapolation to the species likely to be exposed in practice.

Tests used only as part of a research programme must also be reliable and consistent, but they are likely to be more specialized, and standardization is of lesser importance.

The choice of which earthworm species to use for laboratory tests has been controversial. There are practical problems arising from current uncertainties about the taxonomy of earthworms, so that it is not clear whether animals from different sources are comparable, nor whether tests supposedly carried out on the same species are using truly uniform material. Also, the ecological diversity of earthworms creates a problem in choosing representative species. These issues are discussed by Bouché (1992).

There is a contrast between the benefits of a highly standardized approach using a single earthworm such as the *Eisenia fetida* complex* (which does not occur in most soils, but is convenient for testing and for captive breeding, and for which there is a substantial comparative database), and the desire to use species that are present in the field. There is currently no consensus about which, and how many, other species are suitable for laboratory studies beyond the initial screening tests. Within the scope of a stepwise testing approach, several kinds of laboratory tests may have a place.

The substrates used in laboratory tests are diverse, including standard artificial soil, filter paper, natural soils, and other media (Reinecke, 1992). Each has advantages and disadvantages, and changes in the composition of an artificial soil can affect the results. Therefore it is important to describe, and if possible to standardize, the soils used. Although procedures have been proposed which allow a reasonably good extrapolation from artificial soils to natural soils (Van Gestel, 1992), it must be recognised that tests in the laboratory cannot simulate the structure of soil or behaviour of earthworms in the field.

Workshop participants discussed the design of laboratory test protocols in detail, from which the following points emerged.

RECOMMENDATION 3

For a preliminary screening method, it is appropriate to adopt the Artificial Soil Test (OECD 1984) as a standard, using earthworms of the Eisenia fetida *complex* as the test species. Mortality should be the main endpoint,*

* Because of the confusion in earthworm taxonomy, all species of *Eisenia fetida* are referred to in this document as the '*Eisenia fetida* complex'. Details of the taxa which this comprises are given by Bouché (1992).

expressed as a LC_{50} (median lethal concentration) and NOEC (no observed effect concentration). This protocol is suitable for most compounds, but different approaches may be preferable in special cases (e.g. granular products).

RECOMMENDATION 4

For higher-tier laboratory tests, to be carried out later in the sequence, more refined methods are needed, taking account of appropriate routes of exposure and other features of field conditions. The exact design of tests should allow exposure to be modified according to relevant field conditions, to permit tests to be tailored to particular cases. Some general principles which should be followed are outlined in Recommendations 5–9.

RECOMMENDATION 5

The use of juvenile earthworms (for example E. fetida *complex) should be considered for tests in which growth is an endpoint, and where it is intended to examine effects on bodyweight changes without complications due to reproduction.*

RECOMMENDATION 6

In assessing effects on reproduction, both fecundity and fertility must be examined. This can be achieved either by separate measurements of the numbers of cocoons and their hatching, or in combination, by recording the numbers of juveniles produced.

RECOMMENDATION 7

Where possible, use captive bred worms in laboratory tests, so as to assure consistency of age, previous history, etc. Quality standards for the breeding of earthworms should be adopted.

RECOMMENDATION 8

'Voucher specimens' (i.e. typical examples of the animals, documented and preserved for future reference) of the stock used for tests should be kept, to allow confirmation of the material tested (Martin, 1990).

RECOMMENDATION 9

Earthworms should be provided with food in tests, if body-weight changes and reproduction are to be measured.

RECOMMENDATION 10

Encouragement should be given to research and development to further improve the standard artificial soil test protocol, and to incorporate endpoints other than mortality.

RECOMMENDATION 11

Models should be developed to allow reliable extrapolation from standard test conditions to natural soils, both to improve the estimation of likely exposure in the field, and to predict population-level effects.

RECOMMENDATION 12

Develop new methods to identify effects in the laboratory and extrapolate to those occurring in the field, including physiological techniques and biochemical assays.

RECOMMENDATION 13

Initiate a common database in which existing test data can be pooled, and develop expert systems to aid interrogation for comparative analyses and interpretation.

Effects in the field

Tests carried out under field conditions have the advantage of realism, and they allow effects of climate and other variation to be taken into account. They are therefore an important part of the testing sequence, to be carried out if laboratory tests suggest a need.

The species present at a field site, and their abundance, depend on ecological conditions, such as soil structure and drainage. In cultivated land they are also influenced by methods of soil cultivation, fertiliser applications, and plant protection practices.

Because field conditions are highly variable, it is appropriate to have a flexible approach to the design of test protocols. However, greater standardization than has been achieved in the past is necessary, to permit valid comparisons between studies.

Test sites can be chosen according to several features; these include soil characteristics, earthworm density, and the pattern of land-use (e.g. grass or cultivated fields). Often it is difficult to test effects directly on cultivated land, if there is a shortage of sites with sufficiently high earthworm densities. As an alternative, plots on minimally cultivated grassland can provide a high-density model, as a preliminary field test which might be followed by further studies under real use conditions in arable fields.

The size of test plots has varied in past studies. Size is a compromise between the need to ensure sufficient replication to allow proper statistical analysis, and the desire to have realistically large treated areas in which the effects of immigration and emigration are small and samples can be taken periodically without significantly depleting the populations of earthworms.

Methods of sampling earthworms are of two kinds – those which cause the animals to emerge from the soil, and those which involve direct removal by the experimenter. Both can be used, separately or in combination, in order to obtain representative samples of all species. The duration of the test period after treatment, and the frequency with which samples are taken, will depend on the species of earthworms, the chemical and its persistence in soil.

Interpretation of adverse effects at the population level requires more than just a measure of mortality. Measurement of abundance and biomass is essential. Probably few other parameters can be easily measured. Adult:juvenile ratios can also demonstrate changes in reproduction. Ideally, each species should be evaluated separately, but for reasons of practicality it may be necessary to combine information for earthworms of each particular ecological type.

It is helpful to define levels of concern (e.g. a certain reduction in population density), and to ensure that the tests used have the power to detect effects of that size. However, in the interpretation of results, care should be taken to distinguish between the statistical significance of changes and their biological significance, which is currently hard to identify. The implications of changes caused by chemicals must be looked at in terms of the scale of reductions in density and the time taken for recovery. The level of 'natural' fluctuations is also relevant for interpretation. In the absence of detailed knowledge, simple thresholds may be used as acceptable criteria until a better understanding is achieved. Significant reduction in population density (e.g. 30% to 50% below normal levels) and the speed at which numbers or biomass recover to the level in untreated sites, provide separate criteria, either of which may indicate a serious effect. If the principal concern is for soil processes (e.g. decomposition or mineralization), these functions may be measured directly, as well as by assessing earthworm populations.

It is important to explore the relationship between the results of laboratory and field tests, so as to allow useful predictions from the laboratory testing stage. When consistent data from a laboratory method are compared to those from a similarly consistent field approach, the correspondence is quite good (Van Gestel, 1991; Heimbach, 1992). However, generalisation of such comparisons to other situations has to be undertaken cautiously. More such comparisons between field and laboratory would be valuable.

RECOMMENDATION 14

Develop a standard field protocol which can be used for a preliminary field assessment of products that are to be applied to cultivated land. This should include evaluation of the relationship between effects on grassland and cultivated plots. If substantial effects are detected, further testing may be appropriate in realistic application conditions on cultivated land.

RECOMMENDATION 15

The normal application rate should be used in field tests, along with a higher rate (e.g. four times normal) to give information about the dose/effect relationship, and to provide a margin of safety.

RECOMMENDATION 16

Initiate work to determine the most suitable plot size for standard field test protocols, so as to minimise edge effects and allow for the periodic removal of samples. As an interim measure, a size of 10 m × 10 m is recommended.

RECOMMENDATION 17

Tests should include a 'positive control' treatment, (i.e. a well-known toxic standard) and should be adequately replicated to detect the smallest population effects expected.

RECOMMENDATION 18

Encourage efforts to improve knowledge of the ecological significance of effects on earthworm populations. This should include models to define the size of effects that can be tolerated by earthworm populations, taking account of repeated applications of pesticides, and relevant population dynamics.

RECOMMENDATION 19

To allow consistent interpretation until the significance of population level effects is better understood, values of 30% to 50% reduction in population density, and failure to recover to untreated control levels within 12 months can be used as criteria for defining a serious effect in field tests.

Estimating exposure

The first, simple approximation in the estimation of exposure for earthworms is to use the application rate of a product to calculate the amount present per unit of ground area. It is important to know how this figure relates to the concentration of chemical used in laboratory tests, and to the actual exposure experienced by earthworms in the soil.

Understanding the fate of the particular chemical concerned (i.e. its distribution in the soil profile, metabolism, and persistence) can indicate whether it is present at depths that are relevant to earthworms. Some species are near the surface, whereas others burrow deeply and may often be below the depth at which high pesticide concentrations occur, although burrowing earthworms will come to the surface to feed.

The likely quantities available to earthworms can be calculated on the assumptions that soil is evenly packed, that the chemical is evenly distributed, and that all the material is present in an upper soil layer of a certain depth. This critical depth limit might be different in grassland and cultivated sites, for example, and may also depend on the pattern of use of the product. Assessments of risk using such assumptions may be sensitive to the particular soil depth chosen for the calculation, which should therefore be clearly indicated. If feasible, analysis of sensitivity should be incorporated.

The behaviour of earthworms can greatly modify their actual exposure, compared to theoretically estimated figures. Daily and seasonal migrations (vertical and horizontal), and avoidance of toxic chemicals, can reduce the risk by lowering exposure, whereas in certain circumstances exposure can be magnified, if worms pick up residues from the walls of their burrows, for example. Such factors are hard to measure, and have rarely been considered in previous tests. There are also substantial differences between earthworm species, and between adults and juveniles, which are likely to influence effects through differences in exposure.

RECOMMENDATION 20

As a general criterion for the calculation of chemical concentrations in soil relevant to earthworms, the top 5 cm is appropriate, assuming an even distribution of chemical. Sensitivity analysis should be incorporated into calculations based on this assumption.

RECOMMENDATION 21

Behavioural effects which might influence the exposure of earthworms should be taken into account, and methods for observations of behaviour should be developed. Differences between species in behaviour deserve further investigation.

RECOMMENDATION 22

Measurements of chemical residues in earthworms should be considered in cases where direct information on exposure is desired, and where there is a strong likelihood that predators may be poisoned by eating contaminated earthworms.

Significance of residues in earthworms

The nature of the toxic effects on earthworms after a given exposure to a chemical can be examined empirically, demonstrating substantial species-differences in many cases. A different approach is to undertake research to understand the biochemical basis of differences in the pathways of metabolism and detoxication of chemicals. This is a rapidly developing field, in which future progress is likely to bring new insights. Better understanding of the reasons for species-differences may help to improve predictions and extrapolations based on simple tests.

Residues in earthworms, whether live or dead, are a potential hazard to earthworm-eating predators. To assess such risks, it is desirable to measure the levels of residues, rather than to rely on uncertain estimates. Also, the behaviour of the predators is critical in determining their exposure, and an analysis of exposure, including the distribution of worms, is an essential stage.

Examination of secondary poisoning is not necessary in all cases. The 'triggers' which signal the need for such an assessment include a high application rate relative to vertebrate toxicity (in which case there is likely to be potential for direct poisoning), and long-term persistence, together with significant potential for bioaccumulation, indicated by the octanol-water partition coefficient.

RECOMMENDATION 23

Encourage research to develop a reliable basis, founded on biochemical understanding, for predicting differences in sensitivity to chemical poisoning between earthworm species. This will involve studies on the mechanisms of degradation, metabolism and detoxication of chemicals in worms.

RECOMMENDATION 24

Conduct assessments of potential secondary poisoning hazards in cases where the chemical is toxic to earthworm predators, persistent, and has a significant potential for bioaccumulation.

RECOMMENDATION 25

Tests to measure toxicity to an earthworm-eating predator may be appropriate when the properties and use of pesticides indicate a potential risk.

Risk assessment

Most countries operate regulatory schemes to govern the assessment of environmental safety of pesticides and other products. Their aim is to assure a high degree of safety before approval for use is granted, based on prediction from experimental data gathered under laboratory and field conditions. Generally, decisions are guided by whether a particular threshold ratio between toxicity and exposure is exceeded, with an appropriate margin of safety. The use of safety factors is not yet widely agreed. To some extent, the factors selected are inevitably arbitrary, although ideally they should reflect uncertainty in the estimates of toxicity and exposure. For example, knowledge of how much species differ in susceptibility, and variation in chemical concentration could help in the choice of a suitable factor.

Some assessment schemes dictate that all products to be used outdoors should be examined for potential effects on earthworms, whereas others rarely address the issue, or only do so as part of the assessment of risk to vertebrates.

The decision-making procedure proposed for risk assessment provides a link between all the other aspects of earthworm toxicology discussed at the Workshop. Its outcome is usually a classification of chemical uses into those causing 'high', 'intermediate' or 'low' risks, rather than a precise quantification. This has the advantage of easing comparisons between tests and between chemicals, provided that the criteria for putting them into these categories are consistent.

There are several possible test sequences which can provide the data needed for risk assessment. In particular, sublethal effects and reproduction might be examined before or after field tests are conducted. The best approach for a particular case may depend on the nature of the chemical and its toxic effects, and on the habitat in which earthworms are exposed. However, in view of the difficulties in measuring some kinds of effects in the field, it may be generally useful to carry out laboratory tests of these aspects before proceeding to field studies, when an intermediate level of hazard is predicted.

It is important to consider the occasional high risks that may occur in extreme conditions. This can be done by testing in 'worst-case' conditions (e.g. using very sensitive indicator species, or higher than recommended application rates), by employing safety factors, or by extrapolation from experimental exposure data.

RECOMMENDATION 26

Risk assessment should be carried out for each type of land in which a product is used (e.g. orchards, grassland and cultivated land), and should take account of the principal ecological categories of earthworms because of their likely differences in exposure.

RECOMMENDATION 27

Risk assessment schemes should be kept flexible, to allow the special features of particular chemical uses to be taken into account, and should include recognition of the need for expert judgement where appropriate (e.g. in deciding on extrapolation of data from laboratory to field, or assessing the significance of population changes).

RECOMMENDATION 28

Guidelines on risk assessment of plant protection products, such as those being produced by EPPO and the Council of Europe (Greig-Smith, 1992), which include decision-making procedures for effects on earthworms, should be adopted as a framework for individual assessment schemes.

RECOMMENDATION 29

Work should be done to improve the knowledge needed to set safety factors for toxicity:exposure ratios where such factors are appropriate. This involves agreement on the principles behind the choice of factors (i.e. levels of uncertainty in estimates or extrapolations), and also analysis of data on existing approved chemicals, to indicate the uncertainty that occurs in practice.

RECOMMENDATION 30

Agreement should be reached on consistent criteria for placing chemicals into the categories of 'high', 'intermediate', 'low' and 'negligible' risk to earthworms.

RECOMMENDATION 31

Products should be labelled to indicate the level of risk to earthworms, with specified recommendations for use, including measures which are designed to minimise risks in practice.

RECOMMENDATION 32

Assessments should be directed at identifying the risks arising from normal use of products, but should also take account of variations in environmental conditions (e.g. weather) or minor unintended deviations from the proper use, which may cause higher risks.

RECOMMENDATION 33

Attention should be given to potential indirect adverse effects on earthworms, particularly in the case of those herbicides or fungicides which have low toxicity to earthworms. Research is needed to improve knowledge of these effects.

Use of earthworms in bioassays

One aspect which was rather different to the rest of the Workshop's programme concerned the use of earthworms as tools in the detection and measurement of chemical contamination. This involves bioassays, carried out at contaminated sites in the field, or in the laboratory using soil taken from such sites, and is complemented by monitoring of residues.

This approach has several advantages, particularly where there are mixtures of chemicals; it is inexpensive, and may be more sensitive than chemical analysis, and it can be related directly to toxic effects of contamination for the soil fauna. However, bioassays of this kind cannot provide an early warning of problems, and long-term monitoring of earthworm populations may also be appropriate.

The need for standardization of methods used in these bioassays is not as great as in risk assessment studies, although consistent approaches may be necessary if tests are used to determine whether regulatory limits have been breached. Also, there is little need for test species or conditions to simulate natural exposure of worms to the chemical.

RECOMMENDATION 34

Encouragement should be given to the use of earthworms for bioassays, and to develop new sensitive endpoints such as immune functions and other sublethal parameters.

Future action

In order to ensure that progress is made in research, and in the promotion of these recommendations, the organising committee feels that further international collaboration should be fostered by the establishment of a coordinating group concerned with all aspects of earthworm ecotoxicology.

RECOMMENDATION 35

A small coordinating group should be established, to monitor progress in the implementation of these recommendations, and other activities relevant to earthworm ecotoxicology. This will include planning for a further Workshop, in about 5 years time, to review developments.

References

AMMON, H.U. (1985). Worm toxicity tests using *Tubifex tubifex. Les Colloques de l' INRA,* **31**, 303–317.

BIERI, M. (1989). Guidelines for the Daniel funnel test – a laboratory test to measure side effects of pesticides on the earthworm *Lumbricus terrestris* L. IOBC Bulletin (in press).

BOUCHÉ, M.B. (1992). Earthworm species and ecotoxicological studies. In: *Ecotoxicology of Earthworms* (H. Becker *et al.*, eds) pp. 20–35, Intercept, Hants, UK.

DREWES, C.D. AND LINGAMNENI, A. (1992) Use of earthworms in eco-neurotoxicity testing: sublethal effects of carbofuran in *Lumbricus terrestris*. In: *Ecotoxicology of Earthworms* (H. Becker *et al.*, eds) pp. 63–72, Intercept, Hants, UK.

GREIG-SMITH, P.W. (1992). Risk assessment approaches in the UK. for the side effects of pesticides one earthworms. In: *Ecotoxicology of Earthworms* (H. Becker *et al.*, eds) pp 159–168, Intercept, Hants, UK.

HEIMBACH, F. (1992). Effects of pesticides on earthworm populations: comparison of results from laboratory and field tests. In: *Ecotoxicology of Earthworms* (H. Becker *et al.*, eds) pp 100–106, Intercept, Hants, UK.

KOKTA, C. (1992). A laboratory test on sublethal effects of pesticides on *Eisenia fetida*. In: *Ecotoxicology of Earthworms* (H. Becker *et al.*, eds) pp 213–216, Intercept, Hants, UK.

MARTIN, N.A. (1990). Voucher specimens : A way to protect the value of your research. *Biology and Fertility of Soils* **9**, 93–94.

REINECKE, A.J. (1992). A review of ecotoxicological test methods using earthworms. In: *Ecotoxicology of Earthworms*. (H. Becker *et al.*, eds) pp. 7–19, Intercept, Hants, UK.

OECD (1984). Guideline for testing of chemicals No. 207. Earthworm, acute toxicity test. Adopted 4 April 1984. OECD, Paris.

VAN GESTEL, C.A.M., VAN DIS, W.A., VAN BREEMEN, E.M. AND SPARENBURG, P.M. (1989). Development of a standardised reproduction test with the earthworm species *Eisenia fetida andrei* using copper, pentachlorophenol, and 2,4-dichloroaniline. *J. Ecotoxicology and Environmental Safety* **18**, 305–312.

VAN GESTEL, C.A.M. (1991). Validation of earthworm toxicity tests by comparison with field studies: a review on benomyl, carbendazim, carbofuran and carbaryl. *Ecotoxicology and Environmental Safety*, **23**, 221–236.

VAN GESTEL, C.A.M. (1992). The influence of soil characteristics on the toxicity of chemicals for earthworms – a review. In: *Ecotoxicology of Earthworms* (H. Becker *et al.*, eds) pp 44–54, Intercept, Hants, UK.

VAN STRAALEN, N.M. AND DENNEMAN, C.A.J. (1989). Ecotoxicological evaluation of soil quality criteria. *J. Ecotoxicology and Environmental Safety* **18**, 241–251.

Appendix – List of participants

C. Aldridge, MAFF, Harpenden, UK
K. Barrett, Schering Agrochemcials Ltd, Saffron Walden, UK
H. Becker, Biologische Bundesanstalt fur Land- und Forstwirtschaft, Berlin, Germany
A.C. Belfroid, University of Utrecht, Netherlands
N. Beyer, Fish & Wildlife Service, Laurel, Maryland, USA
M. Bieri, Rushchlikon, Switzerland
M.B. Bouché, CEPE/CNRS, Hunt Pellie, France
H. Bouwman, Potchefstroom University, S. Africa
C.A. Callahan, Environmental Protection Agency, Corvallis, Oregon, USA
O. Christiensen, University of Aarhus, Denmark
D. Cluzeau, University of Rennes, Paimpoint, France
A.S. Cooke, English Nature, Peterborough, UK
J.M. Coulson, ICI Agrochemicals, Bracknell, UK
H. Craven, Environmental Protection Agency, Washington DC, USA
G.P. Dohmen, BASF, Limburgerhof, Germany
C.D. Drewes, Iowa State University, Ames, Iowa, USA
D. Ebert, BASF, Limburgerhof, Germany
C.A. Edwards, Ohio State University, Columbus, Ohio, USA
P.J. Edwards, ICI Agrochemcials, Bracknell, UK
H. Eijsackers, Netherlands Integrated Soil Research Programme, Wageningen, Netherlands
L. Fayolle, INRA, Dijon, France
C.L. Fordham, ESE, Eaglewood, Colorado, USA
L.C. Fitzpatrick, University of North Texas, Denton, Texas, USA
A.J. Goven, University of North texas, Denton, Texas, USA
P.W. Greig-Smith, MAFF Central Science Laboratory, Worplesdon, UK
C. Hansen, Danish National Agency of Environmental Protection, Copenhagen, Denmark
A.R. Hardy, MAFF Central Science Laboratory, Slough, UK
E. Harrison, Shell Researech Centre, Sittingbourne, UK
F. Heimbach, Bayer AG, Leverkusen, Germany
R. Heusel, Hoechst AG, Frankfurt, Germany
C.H. Hogger, Federal Research Station for Agronomy, Zurich, Switzerland
M.C. Huet, IRCHA/INERIS, Vert-le-Petit, France
C. Kokta, Biologische Bundesanstalt fur Land- under Forstwirtschaft, Braunschweig, Germany
J.C. Kuehle, Forschungsstelle fur Terrestrische Okotoxikologie und Bioindikation, Bonn, Germany
L.A.M. Kuijpers, Duphar BV, ZG'S-Graveland, Netherlands
H. Kula, TU Braunschweig, Braunschweig, Germany
P.J. Lawlor, Dept. of Agriculture & Food, Dublin, Ireland
A. Lofs, Osterbybruk, Sweden
W-C. Ma, Research Institute for Nature Management, Arnhem, Netherlands

J.M. Marquenie, Nederlands Aardoilie Maatschappig BV, Assen, Netherlands
N.A. Martin, DSIR Plant Protection, Auckland, New Zealand
J.C. Mather, University of Aarhus, Denmark
R.E. Moorcock, Environmental Protection Agency, Washington DC, USA
A.J. Morgan, University of Wales, Cardiff, UK
B. Moulin, Laboratoires de Recherche de L'Oreal, Aulnay-sous-Bois, France
G. Purvis, University College Dublin, Ireland
J. Rauscher, Dallas, Texas, USA
P.B. Rees, Life Science Research Ltd, Eye, UK
A.J. Reinecke, Potchefstroom University, South Africa
B. Roberts, Northeast Louisiana University, Monroe, Louisiana, USA
J. Roembke, Battelle Institute, Frankfurt, Germany
J. Simmers, Corps of Engineers, Vicksburg, Massachusetts, USA
J. Stenersen, University of Oslo, Norway
A.D. Tomlin, Argriculture Canada, London, Ontario, Canada
A. Toufic, Laboratoires de Recherche de L'Oreal, Aulnay-sous-Bois, France
C.A.M. van Gestel, RIVM, Bilthoven, Netherlands
N.M. van Straalen, Vrije Universiteit, Amsterdam, Netherlands
B.J. Venables, TRAC Laboratories Inc., Denton, Texas, USA
A. Vial, Ciba Geigy Ltd, Basel, Switzerland
R. Viswanathan, GSF, Neuherberg, Germany

Index